T0343878

Refractory Technology

This book explains the refractories from different fundamental aspects, even with the support of phase diagrams, and also details the prominent applications of these industrial materials. The initial chapters cover fundamentals of refractories, classifications, properties, and testing, while later chapters describe different common shaped and unshaped refractories in detail and special refractories in a concise manner. The second edition includes new classifications, microstructures, the effect of impurities with binary and ternary phase diagrams, and recent trends in refractories including homework problems and an updated bibliography.

Features:

- Provides exclusive material on refractories.
- Discusses detailed descriptions of different shaped and unshaped refractories.
- Covers concepts like environmental issues, recycling, and nanotechnology.
- Explores details on testing and specifications including thermochemical and corrosion behavior.
- Includes a separate chapter on trends of refractories and related issues.

This book is aimed at junior/senior undergraduate students and researchers of ceramics, metallurgical engineering, and refractories.

Refractory Technology
Fundamentals and Applications

Second Edition

Ritwik Sarkar

CRC Press
Taylor & Francis Group
Boca Raton London New York

CRC Press is an imprint of the
Taylor & Francis Group, an **informa** business

Designed cover image: Ritwik Sarkar

Second edition published 2024
by CRC Press
2385 NW Executive Center Drive, Suite 320, Boca Raton FL 33431

and by CRC Press
4 Park Square, Milton Park, Abingdon, Oxon, OX14 4RN

CRC Press is an imprint of Taylor & Francis Group, LLC

© 2024 Ritwik Sarkar

First edition published by CRC Press 2017

Reasonable efforts have been made to publish reliable data and information, but the author and publisher cannot assume responsibility for the validity of all materials or the consequences of their use. The authors and publishers have attempted to trace the copyright holders of all material reproduced in this publication and apologize to copyright holders if permission to publish in this form has not been obtained. If any copyright material has not been acknowledged please write and let us know so we may rectify in any future reprint.

Except as permitted under U.S. Copyright Law, no part of this book may be reprinted, reproduced, transmitted, or utilized in any form by any electronic, mechanical, or other means, now known or hereafter invented, including photocopying, microfilming, and recording, or in any information storage or retrieval system, without written permission from the publishers.

For permission to photocopy or use material electronically from this work, access www.copyright. com or contact the Copyright Clearance Center, Inc. (CCC), 222 Rosewood Drive, Danvers, MA 01923, 978-750-8400. For works that are not available on CCC please contact mpkbookspermissions@ tandf.co.uk

Trademark notice: Product or corporate names may be trademarks or registered trademarks and are used only for identification and explanation without intent to infringe.

ISBN: 9781032131405 (hbk)
ISBN: 9781032131412 (pbk)
ISBN: 9781003227854 (ebk)

DOI: 10.1201/9781003227854

Typeset in Times
by codeMantra

Dedicated to My
Teachers,
Parents and
Family

Contents

About the Author

Dr. Ritwik Sarkar (b. 1972) is a Professor in the Department of Ceramic Engineering, National Institute of Technology, Rourkela, India since 2009. He has completed his graduation in Ceramic Technology at the University of Calcutta in 1993, postgraduation in Ceramic Engineering at Banaras Hindu University (BHU) in 1995, and a Ph.D. degree at Jadavpur University in 2003, all from India. Before his current profession, Dr. Sarkar has worked as General Manager—Technology, IFGL Refractories Ltd., India during 2008–2009. He has also worked as a Scientist in Central Glass and Ceramic Research Institute, India during 2001–2008, in Research and Consultancy Directorate, ACC Ltd., Thane, India during 1999–2001 and H & R Johnson (I) Ltd., Thane in 1995. He was also a Post-Doctoral Research Fellow in the Institute of Ceramic Components in Mechanical Engineering (IKKM), RWTH, Aachen, Germany, with DAAD (German Academic Exchange Service) Fellowship, during 2003–2004.

The current areas of interest and research works of Dr. Sarkar include the development of refractory aggregates, unshaped and castable refractories, use of nanocarbon in carbon-containing refractories, spinel-based ceramics, machinable bioceramics, and solid-waste utilization. A Life Member of The Indian Ceramic Society and Fellow of the Indian Institute of Ceramics, Dr. Sarkar has worked as the Assistant Editor of *IRMA (Indian Refractory Makers' Association) Journal*, Editorial Board member of the Transactions of the Indian Ceramic Society and is reviewer of many prestigious research journals. Dr. Sarkar has more than 170 research publications and 9 patents to his credit. Dr Sarkar also appears in the list of *World's Top 2% Scientists*, prepared by Stanford University, US.

Dr. Sarkar has received Gold Medal from BHU, Jawaharlal Nehru Memorial Fund's award for his Academic Excellence and Young Scientist Award, Ganpule Award, and Deokaran Award from Indian Ceramic Society for his scientific and research contributions to Ceramic Science and Engineering.

Preface to the Second Edition

I feel overwhelmed by the book for completing more than 5 years of demand in the field of ceramics and refractories and that too as a primary referred and text book on refractories for the benefits of the students, researchers, and the professionals. I take this opportunity to thank the readers, and especially the research and teaching community for providing me valuable feedback and suggestions for further improvement of this book.

This second edition of *Refractory Technology: Fundamentals and Applications* is an updated and improved version of the first edition. The whole book is upgraded with up-to-date information and incorporations. Updated information on refractory uses, steel statistics, and commercial aspects of refractories are provided in the first chapter. The second chapter is enriched with few other concepts of classification of refractories. Detailed information on microstructure of refractories is provided in the properties of refractories. Thermal properties are enriched with equipment figures and detailed description. All the individual refractories, like silica, alumina, magnesia, dolomite, etc., described between Chapters 5 and 10, are discussed in detail for the effects of impurities present in them with the help of relevant ternary phase diagrams. Details of binary and ternary compounds that may form in the system and their melting characters are tabulated for easy understanding. Unshaped refractory in Chapter 13 is enriched with the schematic figures of bond materials. Also, effect of particle size distribution on castables is described in detail. Information and data on the various types of commercially used/available unshaped refractories are tabulated and provided in different tables for easy understanding on the unshaped refractories. Applications of different unshaped refractories are covered in a systematic manner for better understanding. A new chapter, entitled "Trend of refractories and other issues," is included as Chapter 14 that covers the new aspects of refractories like progress in refractories, nanotechnology in refractories, environmental aspects of refractories, and recycling of refractories.

At the end of the preface, I would like to record my sincere thanks to all who have rendered suggestion and help in varieties of ways in making this edition up-to-date and beneficial for the readers.

Ritwik Sarkar

Preface to the First Edition

Ceramic materials have been closely related to the development of human society from its very beginning. Ceramic technology is one of the most ancient technologies, more than 24,000 years old, and at the same time is the most modern, dynamically developing, and diversifying field. The ever-increasing uses of ceramics in different areas of space age technology have made it an area of strategic importance. Among the various classes of ceramics, refractories are the materials having that chemical and physical properties that make them applicable for structures or as components of the system that are exposed to high temperatures.

Use of refractory has started since the invention of fire by humans and its controlled use. In the beginning, people started using clayey mass as refractory to control the fire. In the initial stage of the metal age, crucible/pot-shaped natural rocks were used to soften and sharpen the weapons and primitive tools. With the industrial revolution in the 18th and 19th centuries, concept of refractory technology has also started changing, and new refractory materials other than naturally occurring clayey mass began to be used in different new furnace and kilns. This period can be considered as the starting of the modern refractory technology. Many of the scientific and technological discoveries and advancements would not have taken place, if the progress in refractory did not happen. But for a common man, refractories remained just as a material required for high-temperature processing where metals cannot work.

The operating environment for refractories is becoming increasingly severe with time. The rise in temperatures of operation, greater productivity, rapid thermal fluctuations, extremely corrosive environments, heavier loads, extended service life, and environmental constrictions have demanded superior quality of refractory products. Hence, for every specific application areas, different classes of refractories with specific properties to withstand those environments are essential. Accordingly, various types of refractories have come up with time.

Refractories need to perform in the industry. So it is essential for a refractory engineer to understand the concept of refractory, its basics and fundamentals, raw materials used, manufacturing methods, and properties of each of the refractories used in the different areas of different high-temperature processing. Also, the interactions of the refractories with other materials, present in them as impurities or encountered during use at the user industry, are important. These interactions affect the properties and performance of the refractories. So detailed understanding of these interactions is essential. But, unfortunately, numbers of books available on refractory are extremely less to handle such a vivid industrial product. Hence, an up-to-date book covering these areas is essential to the refractory global community.

This book, *Refractory Technology: Fundamentals and Applications*, has been prepared to meet up such a gap of the scarcity of complete books on refractory and also with an emphasis on fundamentals and application of refractory materials. The first chapter provides a detailed introduction to refractory materials with a historical preview and current status. Chapter 2 deals with the classifications. Different types of classifications used in various refractory and user industries are covered with a brief introduction to each class of refractories. Chapters 3 and 4 deal with the properties of refractories and their testing methods. Properties are also classified as per their nature, like physical, mechanical, thermal, thermomechanical, etc., and each and individual properties are discussed. Details of the testing methods of each of the individual properties of the refractories are described with relevant parameters for better understanding and information. Also, an introduction to standard specifications, as prevailed in different corners of the world, is incorporated in Chapter 4 to have a feeling of actual practice.

Chapters 5–12 deal with the common and individual types of refractories used in various industries. Common items like raw materials, manufacturing methods, subclassifications used in that particular refractory, properties, and applications are there for silica, alumina, fireclay, magnesia, doloma, magnesia–carbon, chrome, etc. refractories. Moreover, the effect of impurities, which these refractories face during their application, has also been discussed with the necessary phase diagrams for better understanding of the refractory in a particular chemical environment. However, binary phase diagrams are only mentioned here as the higher number of components may complicate the understanding. Chapter 12 describes some special refractory materials that are highly useful for the incorporation of certain specific property but used in a limited extent mainly due to their availability and cost.

Unshaped refractory is covered in Chapter 13. As on today, literature and books available on refractories have completely separated these shaped and unshaped materials. In this book, it has been incorporated together to have the idea of unshaped materials in parallel with the shaped ones. Though unshaped refractory has started its commercial success more than 100 years ago, it has strongly developed and been widely used for about last 30 years. Evolving from simple mixes of materials with different sizes, today's unshaped refractory is based on complex and advanced formulations with multiple additives and applied by different advanced techniques resulting in better performances with enhanced life. Chapter 13 includes classification of unshaped refractories, details of various types of materials and additives used in making them, and brief details of each type of unshaped refractories along with the main application areas.

Ritwik Sarkar

1 Introduction to Refractory

1.1 INTRODUCTION

Ceramic materials have been closely related to the development of human society from its very beginning. Ceramic technology is one of the most ancient technologies, more than about 24,000 years old. The earliest-found evidences are from south-central Europe and are mostly sculpted figures. These ceramic articles reveal the societal development and culture of our ancestors, and progress in human civilization. These prehistoric artifacts play a pivotal role in archaeological science for understanding the culture, technology, and behavior of the people in that era. Ceramics are one of the most common artifacts found as fragments of broken pottery, sculpture, toys, ornaments, etc. in an archaeological site.

But at the same time, ceramic is the most modern, dynamically developing, and diversifying field of science. The ever-increasing uses of ceramics in different areas of space age technology have made it a field of strategic importance too. Ceramics as cement, glasses, enamels, porcelains, claywares, etc. have responded to the fundamental human needs by providing building materials for shelter, articles for cooking and storage, and many other aspects. Among the different classes of ceramics, refractories are the materials that are resistant to decomposition or deformation by heat, pressure, or chemical attack, and retain strength even at high temperatures. In other words, refractories are heat-resistant materials that constitute the linings for high-temperature furnaces and reactors and other processing units. They are resistant to thermal stress and other thermal energy-related physical phenomena, and under mechanical loads and shocks, they resist the abrasion and wear of the frictional forces and corrosion by chemical agents. And all these different critical characteristics are required simultaneously in an environment of the different partial pressure of oxygen at high temperatures. Refractories act as the "backbone of industry" for any high-temperature manufacturing sector. They support the production of all the basic and essential commodities manufactured at high temperatures, like iron and steel, aluminum, copper, cement, glass, chemicals and petrochemicals, ceramics, etc.

Hence, it is clear that refractories are essentially required for withstanding the heat, and any process associated with high temperature must require refractories. Table 1.1 shows different high-temperature commercial processes and industries and the associated temperature. All these high-temperature processing industries require refractories. Now, as the temperatures of processing at different industries are varying, obviously the same refractory materials cannot work for all the industries. Hence, as per the conditions prevailed (not only the temperature alone) in an industry or in any high-temperature processing, different types and qualities of refractories are required to apply as per the best suitability, performance, and life.

DOI: 10.1201/9781003227854-1

1

TABLE 1.1
Some High-Temperature Industrial Processes

Industries	Range of Temperatures (°C)
Industrial drying	50–300
Petrochemical industries	100–1100
Hydroxide calcination	400–800
Glass annealing	400–800
Carbon combustion	400–900
Steam boiler	400–1000
Sulfide ore roasting	400–1200
Heat treatment and annealing of metals	500–1300
Aluminum and magnesium	800–1100
Carbonate calcining	800–1300
Sulfate decomposition	800–1400
Foundry industry and rolling mills	900–1400
Salt glazing of conventional ceramics	1000–1300
Fusion process	1000–2200
White ware industries	1100–1500
Glass making	1300–1500
Phosphate decompose	1300–1700
Iron making and steel making	1300–1800
Baking of carbon	1300–1800
Sintering of oxides	1300–1800
Refractories	1300–1850
Portland cement	1350–1700
Sintering of carbides	1500–1900
Sintering of carbide	1500–2000
SiC industries	1800–2200
Refractory metals	1900–2200

1.2 DEFINITION

The word "ceramic" is derived from a Greek word "*keramos*" meaning "potter" or "pottery," and also related to the word "*keramikos*" meaning "of pottery" or "for pottery". Again, "*keramos*," in turn, is related to an old Sanskrit root meaning "to burn" but mainly used to mean "burnt stuff." Again, the word refractory originates from the Latin word "*Refrāctārius*," which means stubborn or obstinate. In the 17th century, it was sometimes spelled as *refractary*, having more touch with its Latin parent but slowly had fallen out of use with time. The word "*refractarius*" is the result of a slight variation in spelling and linked to the Latin verb *refragari*, meaning "to oppose." Again, looking into the meaning of the word refractory as per Cambridge dictionary, the word refractory means "not affected by a treatment, change, or process" and as per Collins dictionary "resistant to heat; hard to melt or work". Again in Webster's English dictionary, refractory is defined as a material that does not significantly deform or change chemically at high temperatures. Also, in other classical dictionaries, refractories are defined as materials that are hard to work with and are especially resistant to heat and pressure.

AS per ASTM C71 specification, refractories are defined as "nonmetallic materials having those chemical and physical properties that make them applicable for structures, or as components of systems, that are exposed to environments above 1,000°F" (~811 K or 538°C). And as per general understanding, refractories are nonmetallic inorganic materials suitable for use at high temperatures in furnace construction. While their primary function is to resist the high temperatures under high load, they are also called on to resist other destructive influences during the hot processing, such as abrasion pressure, chemical attack, and rapid changes in temperature. In practical terms, refractories are the products used for any high-temperature processing to insulate the system, introduce erosion/corrosion/abrasion resistances against the process conditions, and are made mainly from nonmetallic minerals. They are designed and processed in such a way that they become resistant to the corrosive and erosive action of hot gases, liquids, and solids at high temperatures in various types of kilns and furnaces. Refractories provide protection to other substances at high temperature without changing its own character.

Refractories are comprised of a broad class of materials having the above characteristics to varying degrees, for varying periods of time, under varying conditions of application area. There are a wide variety of refractory compositions fabricated in a large range of shapes, sizes, and forms which have been adapted to a wide variety of applications. The common denominator is that whenever a refractory is used it will be subjected to high temperatures.

1.3 BASIC PROPERTY REQUIREMENTS

It is now clear that refractories are the industrial materials that are primarily required for any high-temperature processing and must need the two very basic features, namely, *high-temperature withstand-ability (high melting point) and high strength even at high temperatures*. Other than these two, the refractories need to have very low deformability against time at continuous high-temperature conditions, that is, refractories must have *high creep resistances*. Also, prolonged or repeated use at high temperature may cause some dimensional changes of the refractory due to internal adjustments, which may cause deviation from dimension accuracy and may result in failure of the refractories. Hence, very *high volume stability* even on prolonged use is essentially required.

Again, most of the high-temperature processings are involved with chemical reactions among the various raw materials, additives and chemicals present with varying reactivity (acidity and basicity), and generation of liquids and gaseous reaction products of the processing. Hence, the refractories also have to withstand all such chemical reactions and environments, and need to have *excellent corrosion resistances*. High-temperature processings are again involved with some input materials and produce some desired products. Hence, refractories are encountering the flow of the materials, both charge and product. These materials may be in the state of solid, or liquid, or gas, either in a continuous fashion or in a batch process having one or more entry and exit sides (both the entry of raw materials and exit of reaction products may occur from the same side in a batch type of processing). Hence, refractories are affected by the movement of the materials and their friction, and thus require excellent *resistances against abrasion, erosion, and wear.*

Excellent *mechanical properties*, especially *strength*, at elevated temperatures are necessary for refractories as they have to carry the batch (charge or product) materials even with thrust of fuel and air during processing. Other than this, some basic thermal properties are also important. Increase in the dimension of any materials with increasing temperature is an atomistic property (reversible thermal expansion). Now this increasing dimension of refractory with increasing temperature may cause a problem for the structural integrity of the furnace or the reaction vessel. Different refractories with different expansion properties change the overall dimensions of the refractory lining and may result in a huge thermal strain during heating and may generate crack. Hence, refractory needs to have a *lower thermal expansion* property so that the thermal stress generated will be minimum. Similarly, thermal conductivity is another important fundamental property and is material dependent. The primary purpose of the use of a refractory is to insulate the high-temperature process from low temperature environment. Otherwise, heat will pass through from the process to the environment resulting in an enormous requirement of thermal energy to run the process. Hence, insulating character or *lower thermal conductivity* is preferable for the refractories to reduce the heat loss and make the process economical. But, lower thermal conductivity in a very high-temperature process may produce a very high thermal gradient within the two opposite refractory surfaces: one facing the high temperature of the process and the other facing the ambient atmosphere. This huge temperature gradient will produce significant thermal strain, dimensional difference between these two surfaces, resulting in thermal shock and cracking of the refractory. So, for *high thermal shock resistance*, refractory needs to pass on heat from the inner surface to the outer one, and hence, may require *high thermal conductivity* too. For such situations, multiple layers of refractories are used, and the hot face may be with higher thermal conductivity and the next layer(s) with lower conductivity. Hence, selection of refractory on the basis of thermal conductivity is critical and is dependent on the application demand.

Refractories, as conventional ceramics, are processed from granular/powdered materials which are only sintered at high temperatures (unlike metal/glass product) primarily to develop bond and increase strength. Hence, the gap between the granular mass is present even after sintering, and the presence of porosity (void) is inherent in them. The presence of porosity results in reduced strength and weak resistances to corrosion, abrasion, wear, etc. Hence, *porosity must be minimum for a strong and dense refractory.* Similar to porosity, *permeability* is also detrimental to refractory. But *porosity is desired* for refractories that are used only *for heat insulting purposes* as the pores are filled up with static air, which is one of the best insulators (easily available and economic) and results in a high thermal efficiency for any high-temperature process.

Hence for any application, the requirements from refractories are:

- to withstand high temperatures even under high load
- to have high volume stability

- to sustain under sudden changes of temperatures
- to resist the action of molten metal slag, glass, hot gases, etc.
- to withstand abrasive/wear/erosive forces
- to have a low coefficient of thermal expansion
- to be able to conserve heat
- to not contaminate the product/process material

1.4 HISTORY OF REFRACTORY DEVELOPMENT

The manufacturing process of refractory is quite similar to the formation of basalt, a naturally occurring siliceous rock. Basalt forms from lava that flows out from volcanic eruptions – under the natural geological forces of heat and pressure. Refractory production is somewhat a replication of this natural process where, in general, the naturally occurring (or synthetic) nonmetallic mineral oxides (and some nonoxides in special cases) convert to refractory under the conditions of high heat and pressure, but mostly in solid state.

Since very ancient times, the developments in refractory technology follow the similar track of the developments of the ceramic and metal industries. Refractories are known to exist since the period of Bronze and Iron Ages, more than 10,000 years ago. In the early Bronze Age, a kind of pit kiln was used to fire the earthen wares, where green earthenwares were placed in a hole, dug into the ground for firing. The soil around the dug-out portion used to act a heat-/fire-resistant material, and the upper limit of such kilns was the decomposition temperature of those soil/clayey materials. Naturally, the cavities for the pit kilns were made in such an area where the soil is relatively more heat resistant, insulating, and resistant to any crack formation on heating and breaking. In today's technical term, the refractories that were used for such firing are nothing but clay refractories.

With time, civilization entered into the Iron Age, when remarkable improvements in refractories were observed. Iron was produced by reducing iron ore with charcoal in a furnace at a temperature much higher than the pottery-making temperatures. The furnace lining had to withstand not only the temperature but also the mechanical action and chemical corrosion of the process and wear of the charge and product materials. Natural stones (mostly silica containing), fire clays, a mixture of charcoal and clay were the standard materials for such furnace lining. Evidences of iron-making furnaces were found in many parts of Europe in the 14th century.

Study on the history of glassmaking depicts that the first perfect glass was made in coastal areas of north Syria, Mesopotamia, or Ancient Egypt way back in 3500 BC. In the modern history, evidence of glass pots, the refractory vessel for glass melting, were found in Salem, Massachusetts in 1638. Fire bricks were reported to be manufactured in England in the 17th century and the manufactured bricks were transported to different countries for glass manufacturing. In the late 18th century, fire brick making was started in the US.

Again, if we look into the history of specific refractory making, it can be found that silica refractories were reported to be manufactured initially in South Wales, England way back in 1842, and later in 1899, it was started in the Mount Union region of Pennsylvania, US. Magnesite refractories were initially started in Austria,

Europe around 1880, and its manufacturing was started in Homestead, Pennsylvania, US in 1888. The first-reported application of dolomite lining was found during the invention of the process of phosphorous removal in Bessemer route of steel making by Sidney Gilchrist Thomas and his cousin Percy Gilchrist in 1878. Pure chrome refractories were started since 1896, and chrome–mag refractories were commenced in 1931 when it was found that the mixture of chrome and magnesite had a better tensile strength properties than the components alone and resulted in better thermal shock resistances. Direct bonded chrome–mag refractories were first launched in 1961. In modern days, the official record of the first use of unshaped refractory, as plastic mass, was found in 1914 by W. A. L. Schaefer. Electrofused cast refractories were invented in the mid-1920s, while the inventors, H. Hood and G. S. Fulcher were studying the stubborn, glass-insoluble inclusion "stones."

Thus, the development of refractory is going on with time, and new types or a combination of different refractory components are being studied, invented, and applied for the betterment of the user industry, resulting in a better-quality product, productivity, longevity, and performances of the refractory lining.

1.5 INTERESTING FACTS AND DATA ON REFRACTORIES

Refractories, being the backbone to any high-temperature processing, is a must component for any industry involved with high temperature. But for all the industries, the demand and requirement for refractories and the criticality of applications for the refractories vary depending on the processing parameters, temperatures, chemical environment, production rate, application of the best suitable refractory, etc. Hence, different industries may use different types of refractories as per their best suitability, and performance and life of refractories. Hence, the requirements for refractories by different industries are also varying, and Figure 1.1 shows the global scenario for main refractory-consuming industries with their percentage of consumption to the total refractory produced.

Iron and steel industries are the major consumers of refractories. Iron- and steel-based products are in maximum demand for human civilization compared to any other materials. This is mainly due to the wide availability of iron (Fe is the fourth most abundant element on the earth's crust after oxygen, silicon, and aluminum, with more than 5% crustal abundance), economic, and favorable characteristics. Hence, the consumption of refractories for the iron and steel industries is also high. Hence, as iron and steel industry is the major consumer for refractories, it guides and drives the developments of the refractory industries. It can be seen that the developmental activities and progress in refractories have mostly followed the developments and requirements generated by the advancements in the iron and steel industries.

Again, within the iron and steel industry, there are multiple work stations (unit/plant) where iron is made, processed, converted to steel, steel is processed, molten steel is cast, etc. Each of these work stations uses different high-temperature processes and requires refractories. But the demand, use, and consumption of refractories in different work stations within the iron and steel industry are different. Table 1.2 shows an average percent distribution of refractories consumed at different work stations in iron and steel plants. The table shows that steel ladle, blast furnace

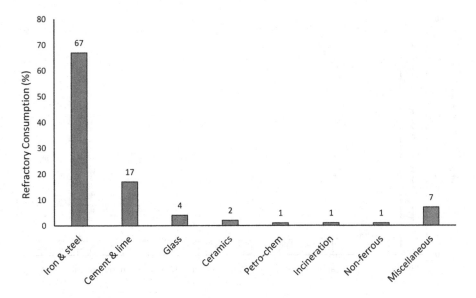

FIGURE 1.1 Industry-wise global refractory consumption.

TABLE 1.2

Percent Distribution of Refractory Consumption in Different Work Stations within Iron and Steel Industry

Name of Work Station/Unit	Refractory Consumed with Respect to Total Consumption in Iron and Steel Industry (%)
Blast furnace and cast house area	21
Hot metal transport (iron ladle/torpedo ladle)	6
Basic oxygen furnace (BOF)/converter	20
Steel teeming ladle	42
Degassing vessels	4
Tundish and continuous casting	7

including cast house area, and basic oxygen furnace (BOF) or converter are the major refractory-consuming units within iron and steel plants. Again, refractories are consumed at different time periods at different workplaces within the same industry. The life of refractory may vary from few minutes to hours to days, months and even up to few years, depending on the prevailing application conditions. Table 1.3 gives some information on the major refractory-consuming areas in some of the user industries, the life of the refractory used there, and the approximate cost involved for refractories with respect to the production cost.

The iron and steel industry is growing with time since long, and global steel production for 2020 was about 1865 million tonnes (Mt). As a country, China takes the

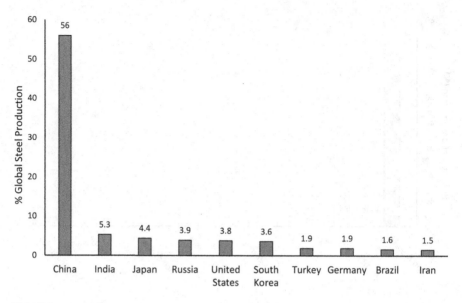

FIGURE 1.2　Major iron- and steel-producing countries.

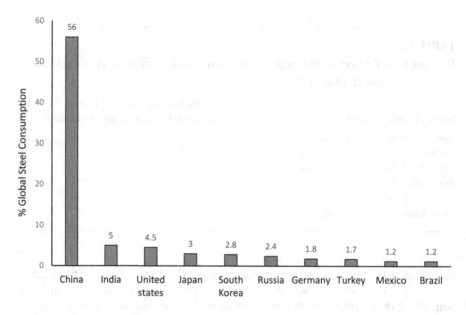

FIGURE 1.3　Major iron- and steel-consuming countries.

lead role in both manufacturing and consuming of iron and steel. Figure 1.2 shows the leading ten steel-producing countries, and it shows that currently China alone produces more than 50% of the global steel manufactured. Also in consumption, China consumes more than 50% of the global steel produced (Figure 1.3). As the refractory industries follow the trend of its consumer industry, that is, the iron and

TABLE 1.3

Average Life and Percent Cost of Refractories in Different Application Areas of User Industries

Industry	Application Area	Gross Average Life of the Refractory	Cost, as Percent Production Cost (%)
Iron and steel industry	Blast furnace	2 days to 10 years	3–4
	Steel converter/ Electric arc furnace/ Steel ladles	30 minutes to 6 months	
	Casting shop	30 minutes to 6 months	
Energy, chemical, environment, petrochemical	Secondary reformer, catalytic cracker, incinerator, furnace	5–10 years	1–2
Cement and lime industry	Rotary kiln	6 months to 1 year	0.5–0.6
Nonferrous industries	Smelter, reverberatory furnace, converter	1–10 years	0.2–0.4
Glass industries	Tank furnace	5–10 years	1–1.5

steel, the largest producer and consumer of iron and steel, China, is the largest producer and consumer of refractories too. And the trend for refractory production and consumption goes very similar to that of the iron and steel; India follows China and so on. Not only for iron and steel, China and India are also the top two cement-manufacturing countries, so the production and consumption of refractories are top for these countries from cement production point of view also. Currently, the global refractory production is about 53 million tons with a gross turnover of more than 23 billion US dollars. Also, globally the refractory market is growing at a compound annual growth rate (CAGR) between 3% and 4%. Figure 1.4 shows the primary refractory manufacturing countries with their percentage of share to the global refractory production.

If we look into the market dynamics for refractories, then we will find that the strength of the refractory industries in the near future is the growing urbanization and industrialization, especially in emerging economies such as China, India, and Brazil. This growth trend has led to significant attention and capital investment in the construction sectors. It is predicted that the global construction market will grow rapidly, and countries like China, India, and the US will account for more than 50% of the global growth till 2030. The increased pace of infrastructure development in the developing countries along with the growth in the construction sector will drive the demand for refractories, especially in the iron and steel, cement, and glass industries. Already, the Asia Pacific region has started showing the trend of increased consumption of refractories in the countries like China, Japan, India, South Korea, and Australia for their iron and steel, power plant, cement, and glass industries in countries. So, the global growth for refractories is going to happen in the coming years.

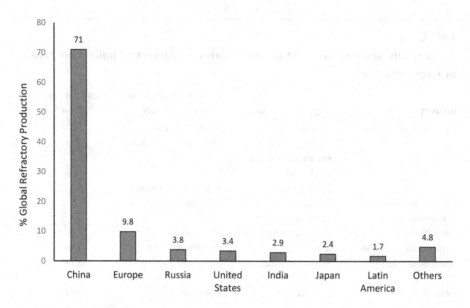

FIGURE 1.4 Major refractory-producing countries.

On the other hand, the hurdles for the growth of refractories in the near future is primarily related to the environment. The release of organic particulate matters (PM) and harmful gases like sulfur dioxide (SO_2), nitrogen oxides (NO_x), carbon monoxide (CO), carbon dioxide (CO_2), fluorides, and volatile organic compounds (VOCs) during the manufacturing of refractories is deadly for human health, and governments are restraining such releases to environment. Release of particulate matters occurs from crushing, grinding, calcining, and drying, and that of volatile organic compounds happen during the heat treatments of organic compounds and binders, namely, tar and pitch. Also, formation of carcinogenic hexavalent chromium in chromium-containing refractories is another life-threatening concern. Manufacturing of refractories, their use, and disposal are under environmental scanning, and governments are putting stricter regulations and restrictions for refractories that will definitely impact the growth. Moreover, due to COVID-19, numerous countries were under lockdown, and the economic and industrial activities were at a halt, which significantly affected the refractories market. The growth rate for refractories has been affected strongly as the production in user industries got affected and became uncertain.

SUMMARY OF THE CHAPTER

Among the various types of ceramics refractories are those materials that have excellent strength even at high temperatures.

Refractories consist of a broad class of materials (mostly oxides) having different types of characteristics to varying degrees at high temperatures that make them capable of withstanding all the different conditions prevailed at the application site.

As per ASTM C71, refractories are "nonmetallic materials having those chemical and physical properties that make them applicable for structures, or as components of systems, that are exposed to environments above 1000°F" (811 K; 538°C).

The basic properties that a refractory must have are

- withstand high temperature
- high strength at high temperatures
- high creep resistance
- high volume stability
- excellent corrosion resistance
- resistance to wear, abrasion, and erosion
- low thermal expansion
- high thermal shock resistance
- thermal conductivity (low or high depending on properties of thermal shock and insulation)
- density and porosity (depending on dense or insulating character requirements)

Refractory is a ancient technology, known to mankind since the invention of fire. It is not a directly consumable product, but rather, it is an ancillary industry. Refractories are a must for all the high-temperature operations. The industry grows along with the growth and requirement of the user industries.

About 70% of the total refractory produced is used in iron and steel industry. The main steel manufacturing and consuming country, China, is the leading manufacturer of refractories.

QUESTIONS AND ASSIGNMENTS

1. What is refractory?
2. What are the basic functions of a refractory material?
3. What are the main properties that a refractory must have?
4. Describe the functions of refractories required for high-temperature processing industries?
5. How the growth of the refractory industry has taken place?

BIBLIOGRAPHY

1. Solomon Musikant, *What Every Engineer Should Know About Ceramics, Preface*, Marcel Dekker Inc., New York, 1991.
2. W. W. Perkins, *Ceramics Glossary 1984*, The American Ceramic Society, Ohio, US, p. 91.
3. James G. Hemrick, H. Wayne Hayden, Peter Angelini, Robert E. Moore, and William L. Headrick, *Refractories for Industrial Processing: Opportunities for Improved Energy Efficiency*, Prepared for the DOE-EERE Industrial Technologies Program, Tennessee, US, 2005.
4. https://www.eere.energy.gov/manufacturing/industriestechnologies/imf/pdfs/refractoriesreportfinal.pdf.
5. Jessica Roberts, Outlook for refractory end markets to 2020, *Presented at the 57th International Colloquium on Refractories*, 24–25 September, Aachen, Germany, 2014.

6. K. Sugita, Historical overview of refractory technology in the steel industry, Nippon Steel Technical Report, No 98, July 2008, pp. 8–17. http://www.nssmc.com/en/tech/report/nsc/pdf/n9803.pdf.

7. D. Tolegenov and D. Tolegenova, Overview of achievements in the refractory industry, *Journal of Economics and Social Sciences* (2020). http://earchive.tpu.ru/bitstream/11683/62044/1/jess-368.pdf.

8. Ahmed Hassan Al-Shorman, *Doctoral Thesis, Entitled, Refractory Ceramic through the Ages: an Archaeometric Study on Finds from Fenan, Jordan and other Sites*, Ruhr University, Bochum, 2009.

9. https://worldsteel.org/wp-content/uploads/ 2021-World-Steel-in-Figures.pdf.

10. *World Steel in Figures*, World Steel Association, Brussels, Belgium, 2018 (ISBN No. 9782930069890).

11. Jorge Madias, A review on recycling of refractories for the iron and steel industry, *Proceedings of the 15th Biennial Unified International Technical Conference on Refractories (UNITECR 2017)*, 26–29 September, Santiago, Chile, 2017.

12. D. A. Jarvis, *Refractories in the British-isles, Refractories World Forum*, 2019. https://www.refractories-worldforum.com/market-news?page=1&news_id=10354&news_title=Refractories+in+the+British+Isles&page=1.

2 Classifications of Refractories

2.1 INTRODUCTION

The wide variety of high-temperature processing conditions across various industries demand diversity in the refractory materials. The diversity in requirements arise from application temperatures, atmospheric conditions that prevail during the processing, mechanical, thermal, chemical, abrading, and wear conditions, etc. In fact, many of the refractory materials have been developed to meet precisely certain service conditions of particular pyroprocessing conditions. To meet this wide variety of application conditions, different classes of refractory materials are required and being developed as the time progresses. For example, the atmosphere of the processing condition demands a refractory of basic or acidic character, the temperature may be very high or low, some applications may require very strong and dense refractory, and, in some cases, the requirement is to insulate the process, etc.

This wide variety of application conditions in different industries are required to be well understood by the refractory manufacturers and developers, and they need to provide a variety of refractories suitable for each and every condition. This variety of refractories can be classified as per various groups/classes based on certain specific categories or properties. The classification of refractories with their detailed sub-classes are described in this chapter. These different classes are based on:

1. Chemical nature: As per chemical nature refractories are classified as:
   a. Acidic refractory
   b. Basic refractory
   c. Neutral refractory
2. Classification based on main constituent and purity: Main constituent: Refractories are commonly classified as per the principal chemical component (normally oxide) present in them and named accordingly as:
   a. Silica refractory: when the main constituent is silica
   b. Alumina refractory: when the main constituent is alumina, etc.
      Within the classification of the main constituent, depending on the percent (purity) of that constituent present, the refractories are further classified as:
   c. 60% alumina brick (means major constituent of the brick is alumina, and the amount of alumina present is 60%)
   d. 70% alumina brick, etc.
3. Manufacturing method: The refractories are manufactured through different techniques or processing, and as per that specific/special techniques the refractories are classified as:

DOI: 10.1201/9781003227854-2

   a. Pressed and fired refractory
   b. Fused cast refractory
   c. Hand molded refractory, etc.
4. Physical form/shape: Classification is also important as per the physical form of the refractory, and the main subclasses are:
   a. Shaped refractory like bricks having a definite shape and size during manufacturing and supply to the user industry
   b. Unshaped refractory like castables, which are well mixed, granular, unfired masses, and do not have any specific shape and dimension during manufacturing and supply to the user industry.
5. Heat duty (application temperature): This classification is based on the maximum temperature that a refractory can withstand or indicates the maximum application temperature. The traditional nomenclatures are:
   a. High heat duty
   b. Medium heat duty
   c. Low heat duty
6. Porosity (insulating nature): Amount of porosity describes the dense (strong) or insulating character of a material. Refractories are classified as per porosity as:
   a. Dense refractory
   b. Insulating refractory
7. Application area: Refractories are also classified as per their application industries, like refractories for iron and steel industry, refractories for glass, etc., even within a particular industries, refractories are subclassed as per the specific area or sop where it is used, For example, for iron and steel industries there are classes like refractories for converter, ladle refractories, tundish refractories, etc.
8. Special refractory: Refractories that do not come under the above broad classifications and are typically used at a much reduced volume, only in special cases, like
   a. Nonoxide-containing refractory
   b. Zirconia refractory

2.2 CLASSIFICATION BASED ON CHEMICAL NATURE

Refractories are very commonly classified based on their chemical behavior, that is, how a refractory will behave in a particular environment. Chemical environments prevailed in the furnace or the processing conditions determine the most suitable refractory for such an applications. As per the chemical nature (affinity) of any material, refractories are classified as acidic, basic, and neutral.

2.2.1 ACIDIC REFRACTORIES

Acidic refractories are those that are resistant to any acidic conditions like, slag, fumes, and gases at high temperatures. But, they are readily attacked by any basic slag or environment. In the presence of any basic materials or environment, the acidic refractories react rapidly, causing a massive corrosion in the refractory lining and resulting in a very poor life of the lining. These refractories are only used in areas where slag and atmosphere are acidic in nature. Examples of acid refractories are:

1. Silica (SiO_2)
2. Fireclay

Both the refractories contain silica as the main material, and, in any basic environment at high temperature, say in the steel ladles or burning zone of cement kilns, they will react and form various alkali and alkaline silicates. These silicates have low melting point, and they further react with solid refractory structure, causing wear of refractory and drastic deterioration of refractory lining. Hence, any acidic refractory performs best in an acidic environment, for example, in glass-melting tank applications, if other application criteria are met.

2.2.2 NEUTRAL REFRACTORIES

Neutral refractories are chemically stable to both acids and bases, and are used in areas where slag and atmosphere are either acidic or basic. Hence, these types of refractories are most commonly preferred. But many of the refractories that behave as neutral at low temperatures behave with some chemical affinity at high temperatures. Hence neutral refractory useful for very high aggressive environments at high temperature is rare.

The typical examples of these materials are:

1. Carbon or graphite (most inert)
2. Alumina (Al_2O_3)
3. Chromites (Cr_2O_3)

Out of this, graphite is the least reactive and is extensively used in metallurgical furnaces where the process of oxidation of carbon can be controlled. Alumina and chromites refractories are stable and neutral in character at low temperatures, but they behave little acidic in nature and react with very strong basic materials at high temperatures. Hence, their chemical neutrality is no longer valid at high temperatures.

2.2.3 BASIC REFRACTORIES

Basic refractories are those materials that are attacked by acidic components but are stable against alkaline slags, dust, fumes, and environments at elevated temperatures. Since these refractories do not react with alkaline slags, they are of considerable importance in basic steel-making processes, nonferrous metallurgical operations, and cement industries. The most common important basic refractories are:

1. Magnesia (MgO)
2. Doloma (CaO.MgO)

Basic refractories are never used in acidic conditions, for example, in glass-melting tank, where they will be washed away very fast by forming low-melting silicates.

2.3 CLASSIFICATION BASED ON MAIN CONSTITUENT AND PURITY

One of the easiest way and commonly practiced classification for refractories is to mention both the main constituents present and its percentage. The major classification is based on the main constituent present and termed as alumina refractory, silica refractory, dolomite refractory, etc., meaning that the refractories are containing alumina, silica, dolomite, etc. as the main constituents. Industrially and commercially, this is the most widely used practice to classify the refractories. Also, for further understanding, the nomenclature also indicates the amount (percentage/purity) of the main constituent, like 90% alumina refractory, 96% silica refractory, etc., meaning that the refractories contain 90% of alumina, 96% of silica in it, respectively. Accordingly, from the purity (amount of impurity) and main constituent, one can also have some idea about the characteristics of the refractory, like, whether it will be acidic or basic, or how the same may work and behave in a particular environment.

2.4 CLASSIFICATION BASED ON MANUFACTURING METHOD

Manufacturing methods of the refractories vary to attain definite shape, size, and specific properties. The nomenclature of the refractories is also done as per the different manufacturing techniques employed. The various subclasses of this classification are as follows.

2.4.1 PRESSED AND FIRED (SINTERED)

This is the most common/conventional type of refractory that is generally found. In this method, particles of different sizes (of various raw materials to get the proper refractory composition) are mixed, and then pressed and fired to attain the desired shape, size, and characteristics. Pressing pressure and firing temperatures are critical parameters, and are fixed as per the constituents of the composition and targeted properties. Use of lower pressure and temperature results in a low strength and relatively porous (less dense) product.

2.4.2 FUSED CAST

Fused cast refractories are important for many industries like glass, iron and steel, and aluminum, etc., where the refractories are subjected to remain in contact with the liquid phase at high temperature for a prolonged period. They are also necessary for petrochemical and other related industries where the refractories are subject to high wear/abrasion/erosion and chemical attack. Fused cast refractories are manufactured by melting mixtures of oxide powders (for the desired composition) in an electric arc furnace (using carbon electrodes) at a temperature exceeding 2000°C and casting the melt in molds, wherein it solidifies through annealing to reduce the generation of strain. These refractories are treated with oxygen while in the molten state to convert all the constituents to their highest oxidized state. A contraction cavity is formed beneath the casting scar during cooling, and so the opposite surface of the casting scar is used as the working face of the refractory.

Compared to conventional refractories, fused cast products have a very dense and highly durable structure using stable mineral substances, and it shows particularly superior mechanical and chemical properties. Also, fusion and slow cooling produce large crystal sizes that reduce surface area for any chemical reaction and resulting in increased corrosion resistances. Low porosities restrict the ingress of corrosive liquids, and the smooth surface prevents the adhesion of slag. These refractories can resist high surface loads due to high hot compression strength. Again, high thermal conductivity from dense structure results in a uniform heat distribution within the refractory, producing an uniform thermal condition. But, as the process involved melting of refractory materials, enormous amount of heat is required to manufacture these refractories that affect the economy.

2.4.3 Hand Molded

These refractories are important for their critical shapes and sizes. Nonconventional, critical shapes and dimensions in refractories are important for many special applications, like in coke ovens, glass tank furnace, etc. Pressing is a simple method for great productivity; but pressing molds cannot produce intricate shapes, as the releasing of the pressed shapes needs a simple design for higher productivity. Also, dimensions of the shape is also limited to pressing process, as much bigger dimensions requires huge pressing capacity, which has certain technical limitations. Hence to manufacture a complicated shape or a large-sized refractory product before firing, hand molding (using techniques like hammering) is the easiest and economic process.

As hand-generated pressure is much lower than the pressing pressure, the amount of water used is about 2–3 times more than that require for pressing. Also, the green strength is lower. Hence, handling and drying of these types of refractories are relatively critical. Higher moisture content and less compaction pressure during shaping of these products produce relatively porous structure and high shrinkage on firing, and so the mold size needs to be adjusted.

2.4.4 Bonding

Most of the refractories are sintered; that means, they have a ceramic bond that has developed during firing in between the loose starting particles from which they have been prepared. Other than this sintering and ceramic bond, many types of bonds are also used for the refractories to develop strength as per the composition and requirement.

Many of the refractories are chemically bonded. That means a different chemical bonding material is used, which is entirely different from the basic composition of the refractory, to create a secondary (chemical) bond between the particles to retain the shape and size, and generate strength and other properties. Mostly, chemical bonding is used for those refractory systems that require the very high temperature to develop a direct sintering bond. The addition of a secondary phase may form a liquid (or reacts and melts) and creates a bonding between the particles of the primary refractory particles. But, the high-temperature properties of the refractory may get affected.

Also, tar/pitch/resin bonding is common for refractories containing carbon in the composition at a considerable amount. Carbon particles, which are having very strong atomic bonding in-between themselves and are very fine, when present in between the refractory oxide particles, do not allow the refractory for mass transfer and sintering even at high temperatures. Hence for carbon bonding, especially, liquid carbonaceous materials are used as bonding material that can polymerize and form a three-dimensional carbon-based network structure, and can hold the whole refractory structure/particles, even at high temperatures, and provide strength.

2.5 CLASSIFICATION BASED ON PHYSICAL FORM OR SHAPE

As per the physical form of the refractory, they are classified as per shaped and unshaped refractories.

2.5.1 SHAPED

These types of refractories are having fixed size and shapes. Shaped refractories are the most common and conventional ones. Shapes may be the standard ones and of special types. Standard shapes are those products whose shapes and dimensions are common, and accepted by most of the refractory manufacturers and users. Standard shaped bricks have dimensions that are conformed to by most refractory manufacturers and are applicable to kilns and furnaces of the same type. Standard shapes are usually required in vast quantities to construct a kiln or furnace.

On the other hand, special shapes are specially made products required for particular kilns or furnaces, with very specific shape and dimensions, and are generally required only in limited quantities. Standard shaped refractories are always machine pressed and thus have uniformity in properties. Special shapes are usually hand molded and are generally associated with slight variation in properties.

2.5.2 UNSHAPED

Unshaped refractories are those that are in loose granular condition, and do not have any particular shape and size while they are transported to the user industry. The shape and size are given as per the requirements of the particular application by making it a flowable/shapeable mass by mixing with liquid (mostly water) and applying the same to the required area by casting, ramming, troweling, gunning, etc. As a very big and single structure can be made from these materials, they are also called monolithic (in Latin, *mono* means single and *lithus* means structure). The formed refractory is fired in its application site, gets sintered, and attains strength with other properties.

2.6 CLASSIFICATION BASED ON HEAT DUTY

Classification of refractories is also done based on the criteria of application temperature, that is, up to what temperature the refractory can perform its duty. This classification is common used in earlier days and presently still valid for low-temperature applicable refractories, especially for fireclay refractories. The primary parameter for this classification is refractoriness or the softening temperature that indicates the maximum temperature the refractory can withstand without deforming. Refractories are classified under four categories in this classification.

2.6.1 LOW HEAT DUTY

These are low-temperature-withstanding refractories having refractoriness up to 1630°C and have pyrometric cone equivalent (PCE) value up to 28. Common example for these refractories are fireclay refractories, low-silica refractories, etc.

2.6.2 INTERMEDIATE (OR MEDIUM) HEAT DUTY

These refractories are capable to withstand higher temperatures than the previous class and have refractoriness in the range of 1630°C–1670°C. The PCE value varies between 28 and 30. Common examples of these refractories are fire clay, low-alumina refractories.

2.6.3 HIGH HEAT DUTY

These refractories have still higher-temperature-withstanding capacity, have refractoriness between 1670°C and 1730°C, and have PCE values between 30 and 33. Example of these refractories is alumina, chromite refractories.

2.6.4 SUPER HEAT DUTY

Among the different types of refractories, this class withstands the highest temperature and has refractoriness greater than 1730°C and PCE value greater than 33. Common examples of these refractories are high alumina, magnesite refractories.

2.7 CLASSIFICATION BASED ON POROSITY (INSULATING) OR THERMAL CONDUCTIVITY

Porosity differentiates the refractories in two major classes: dense and insulating. The same classification is based on thermal conductivity too: dense ones are having higher conductivity, and insulating ones (porous) are having low conductivity due to the presence of pores filled with air, which is a bad conductor of heat.

2.7.1 DENSE

Dense refractories are those that have a very low porosity and are highly sintered to attain the maximum possible packing and densification. Higher densification (lower porosity) comes from more and more contact in-between the particles due to a greater extent of packing (compaction/pressing) and sintering. Greater degree of grain-to-grain contact and sintering will produce greater strength. Hence, the dense refractory shows improved mechanical properties, higher resistances against corrosion, abrasion, wear and erosion, increased thermal conductivity, etc. Conventionally, refractories having porosity below 45% are termed as dense refractories.

2.7.2 INSULATING

Any high-temperature processing requires a significant amount of thermal energy. And, in most of the cases, the energy requirement for the actual process is much lower than the total energy consumed. This is due to the enormous amount of energy loss from the high-temperature processing condition mainly through the walls of the processing container (that is the refractories) into the atmosphere. And, the higher the temperature of processing, the more the chance of heat loss. To reduce the escape of energy from the process, a special kind of refractory lining material for the processing vessel is required, which is called insulating refractory. The primary function of the insulating refractory is to prevent or reduce the rate of heat flow (heat loss) through the walls of the furnaces. These refractories need to withstand the high temperature of its application and simultaneously need to be insulating (low thermal conductivity). Such requirements are achieved by incorporation of porosity in the refractory body, which are actually small air pockets. Air is having very low heat conductivity; hence, the more the porosity, the more the insulating character of the refractory. Generally, refractories having a porosity more than 45% are called as insulating refractories. Small and uniformly distributed pores with a higher amount of porosity result in better quality of insulating character.

2.8 CLASSIFICATION BASED ON APPLICATION AREA

Refractories are applied products that are used in different industries as per the requirement of that industry and particular application area. So, commonly the refractories are also classified as per their application area and grossly termed as per the particular application. Refractories are majorly consumed by the iron and steel industry, so the refractories used for making iron and steel are classified as refractories for iron and steel. Again, as per the diversity in the operating conditions in the iron and steel industry, different types of refractories are used in different shops and metallurgical vessels or processes. Each production shop may require special shapes as well as different technical specifications for refractories that are necessary to meet the process requirements for that shop. Hence, the refractories are often named after the shop/process names. So, within the class of "refractories for iron and steel," there may be subclassification as (i) coke oven refractories, (ii) blast furnace refractories, (iii) steel-making refractories, (iv) ladle refractories, (v) tundish refractories, (vi) calcining plant refractories, (vii) reheating furnace refractories, etc.

Similar type of classification is also there, as per the industries like cement, glass, aluminum, copper, petroleum, chemical, etc. Also, subclassification under each class is also there, as per the name of the processes and shops.

SUMMARY OF THE CHAPTER

Classification of refractories is done as per various parameters; most common of them are chemical nature, manufacturing method, physical form, porosity, heat duty, main constituent, purity, application area, etc.

As per chemical nature, they are acidic (silica), basic (magnesia), and neutral (alumina) types.

As per the manufacturing methods, refractories are classified as pressed and sintered, fused cast, hand molded, chemically bonded, etc.

According to physical form, the refractories are shaped and unshaped (monolithic) types.

As per heat duty, refractory types are super heat duty, high heat duty, medium heat duty, and low heat duty.

As per the main constituents, refractories are classified as silica refractories, alumina refractories, fireclay refractories, magnesia refractories, etc. In the commercial world and manufacturing, mostly refractories are classified as per this classification.

Classification based on purity of the main constituent (90% or 80%) is a subclass of the above types.

As per porosity, refractories are dense (porosity <45%) and insulating (porosity >45%) types.

Refractories are also classified as per the area of application, like refractories for iron and steel industry, refractories for cement industry, etc.

QUESTIONS AND ASSIGNMENTS

1. Why do we need a classification of refractories?
2. What do you understand by classification of refractories?
3. What are the different classification parameters of refractories?
4. Describe in detail the classification of refractories as per chemical nature.
5. Describe in detail about the classification of refractories based on manufacturing methods.
6. Describe the porosity-based classification of refractories.

BIBLIOGRAPHY

1. C. A. Schacht, *Refractories Handbook*, CRC Press, Boca Raton, US, 2004.
2. J. H. Chesters, *Refractories- Production and Properties*, Woodhead Publishing Ltd, Cambridge, 2006.
3. P. P. Budnikov, *The Technology of Ceramics and Refractories*, Translated by Scripta Technica and Edward Arnold, The MIT Press, 4th Ed, 2003.
4. *Harbison-Walker Handbook of Refractory Practice*, Harbison-Walker Refractories Company, Moon Township, PA, 2005.

5. C. Barry Carter and M Grant Norton, *Ceramic Materials: Science and Engineering*, Springer Science Business Media, New York, NY, 2013.
6. Stephen C. Carniglia and Gordon L. Barna, *Handbook of Industrial Refractories Technology: Principles, Types, Properties, and Applications*, Noyes Publications, Saddle River, NJ, 1992.
7. A. Rashid Chesti, *Refractories: Manufacture, Properties and Applications*, Prentice-Hall of India, New Delhi, 1986.

3 Idea of Properties

3.1 INTRODUCTION

Understanding the properties of refractory materials is important for gaining knowledge on the subject that will help for the selection of refractories for a particular application. It also helps in developmental activities, improvement in qualities, and quality control of refractories. Refractories, by definition, are those materials that can resist heat, corrosion, abrasion, and thermal shock and can withstand different degrees of mechanical stress and strain at various temperatures. The compositional adjustments are done for the design of different refractories to optimize the properties that are appropriate for their applications in particular environments.

The quality of any refractory and its suitability for any specific application environment does not depend on any specific property. Rather, it is a combination of different properties that finally decides whether that particular refractory is suitable for any specific environment or not. Hence, it's a group of properties that is important for a refractory. The group of property required for a refractory may change from applications to applications, and the refractories that satisfy all the required property criteria for a specific application are the most suitable ones for such applications.

There are different groups of properties important for any refractory, and the most common groups and their detailed individual descriptions are given in this chapter. Other than the common types of properties like physical, mechanical, thermal, etc., certain specific properties are measured only for refractories. These refractory-specific properties are also described individually at the end of the chapter.

3.2 PHYSICAL PROPERTIES

Properties that are easily measurable, whose values represent the state of a physical system, and the measurement processes do not change the identity of the material are called physical properties. In the measurement of physical properties, there must not be any chemical change, mechanical breakdown of the sample and the physical form will remain intact. Physical properties can be used to characterize mixtures as well as pure substances. The changes in the physical properties of a system can be used to describe its transformations or evolutions between its momentary states.

Hence, by definition, physical property is a characteristic of a substance that can be observed or measured without changing its identity. All of the senses can be used to observe physical properties like color, shape, size, etc. Mass, volume, and density are physical properties. Changing the mass or volume of a substance does not change the substance's identity. The state of matter that describes the physical form of the matter is also a physical property.

For refractories, the main physical properties measured are as follows.

DOI: 10.1201/9781003227854-3

3.2.1 APPARENT POROSITY (AP), TOTAL POROSITY, AND BULK DENSITY (BD)

The primary difference in manufacturing a metal or glass product and a ceramic product is their processing technique. Any metal or glass item is made from its liquid state, and processed to a solid state with specific shape and size. Formation of the liquid phase is beneficial, as it is free from any air or void space (porosity) inside it, unless and otherwise is entrapped during processing. But refractories and ceramics are made up from loose granular mass having different sized particles and then firing it to a high temperature without or with a little liquid phase formation. Particularly for refractories, the liquid phase formation is nearly negligible (other than fused cast products), as that liquid phase will limit the high-temperature properties and applications of the refractory by a drastic reduction in hot strength.

Hence, refractories are not free from air pockets/voids, which in a solid mass is technically termed as "porosity." The porosity of a material is defined as the ratio of its pore volume to its bulk volume. The amount, size, and distribution of porosity control many of the refractory properties and also dictate the suitability of that refractory for a targeted application. Hence, it is important to measure the amount of porosity present in any refractory.

Porosities can be of two different types: open porosity and closed porosity. The porosity that is present on the surface of a refractory is open to the environment called open or surface porosity. As the refractory surfaces are facing all the criticality of the application environment, the porosity present in the surface are most important, and their determination is essential. These surface porosities are technically termed as "apparent porosity." The apparent porosity is the ratio of the volume of the open (surface) pores, into which a liquid can penetrate, to that of the total volume of the sample, expressed as a percentage. This is an important property, especially for the cases where the refractory is in contact with the molten charge and slag. A low apparent porosity is desirable, since it would prevent easy penetration of any liquid in the refractory. Again, pore size and continuity of pores are important, as they influence the behavior of the refractory against the molten material. Connectivity of pores is dangerous, as it helps the liquid to penetrate into the interior of the refractory and cause the drastic deterioration. A large number of small pores is preferable compared to an equivalent volume of large pores from strength and chemical attack points of view. Apparent porosity (AP) is calculated as per the formula,

$$AP \text{ in } \% = (W - D) \cdot 100/(W - S),$$

where D = weight of the dried sample, S = suspended weight of the sample when immersed in liquid, and W = soaked weight, that is, the weight of the sample containing liquid in the surface pores/open pores but not on the free surfaces.

Closed porosities are those pores that are not visible and not accessible unless the sample is broken. There is no direct method to view and measure them. They are closed from all the sides and very important for the thermal conductivity, strength, bulk weight of the samples, etc. Summation of closed and open porosities gives the value of total porosity.

Increase in porosity results in

1. Poor conductivity: Pores or voids spaces are filled up with static air, and air has very low thermal conductivity. Hence, the overall conductivity decreases. A refractory behaves like an insulating material at a very high level of porosity values.
2. Higher resistance against thermal fluctuation: As the pores are open space, they can accommodate the sudden expansions resulting in from the thermal expansion property of sudden thermal fluctuations. Hence, strain generation will be less.
3. Poor strength: Higher porosity means less number of particles in contact, hence less resistance against any load and poor strength.
4. Poor resistance to abrasion, erosion, and wear.
5. Poor resistance to any chemical attack, as the area for any reaction will increase due to the increased pore surfaces, and easy penetration of corrosive liquids and gases.
6. High permeability: Higher porosity increases the chance of pore coalescence and pore connectivity, resulting in higher permeability for any fluid through refractory.

Density is one of the most common and fundamental physical properties of any material. It is defined as the ratio of any shape's mass to its volume. As most of the designs are limited by either their size and or weight, density is an important consideration in many calculations. Now, the refractories have porosity within it, so the density is lower than their actual material's density. The presence of air pockets or voids or pores reduces the mass of the refractory. This reduced mass results in decreased density of the refractory. The reduced density is required to be measured to get the idea of the total amount (weight) of refractory needed for a particular lining volume and overall weight calculation of the lining required to make the furnace and for heat balance calculation. This reduced density value that is useful for the refractories, containing some amount pores in it, is called "bulk density." It is defined as the weight per unit bulk volume of refractory. This bulk volume considers the volume of the sample without the volume of the surface/open pores. This bulk density indicates the actual densification of the material, and gives an idea of the strength development and other related properties. Obviously, this density is lower than the true density of the material, and, the higher the amount of porosity, the lower will be the bulk density values. Bulk density (BD) is calculated as per the formula,

$$BD = D \cdot \rho / (W - S),$$

where D = weight of the dried sample, S = suspended weight of the sample when immersed in liquid, W = soaked weight, that is, the weight of the sample containing liquid in the surface/open pores but not in the free surfaces, and ρ = density of the liquid used for immersion at the test temperature.

3.2.2 SPECIFIC GRAVITY

The specific gravity is the ratio between the ideal (or true) density of an object (without any pore) and that of a reference substance at the same test conditions. Usually, the reference substance is water. In simple concept, specific gravity values indicate whether an object will sink or float or compare a heavier and lighter object. The specific gravity has *no unit*, because it is the ratio between the true density of the test sample and that of the reference sample (water). Specific gravity varies with temperature and pressure, and so both the reference and test samples are to be compared at the same temperature and pressure conditions or otherwise, the specific gravity values are to be corrected for a standard reference temperature and pressure. Specific gravity and bulk density values will match when the total porosity value is "zero", indicating full densification of the sample. For refractories, materials with very high specific gravity are not to be preferred, as high specific gravity means higher amount of material for a fixed volume; so consumption of refractory material for lining a fixed dimension will be higher, and total load (both weight and thermal load) of the lining will be higher with increased cost. Specific gravity values of some common refractory materials are given in Table 3.1.

3.2.3 FIRING SHRINKAGE

This is the dimensional change that a refractory shape may have due to firing. As this property evaluation requires only the measurement of the dimensions of the sample before and after firing, it is considered as a physical property. Shrinkage value is dependent on many parameters, the important ones are, composition, firing temperature, soaking (dwelling) time, shaping pressure, etc. This property is more important for the refractory manufacturer than that of the user, as the manufacturer needs to adjust the unfired dimension and mold dimensions (as per the shrinkage value) to get the desired fired dimensions. Shrinkage, which is dimensional contraction during heat treatment (also during drying), occurs mainly due to removal (and reduction) of the pores (and pore volume) from the sample. Very high shrinkage values are not desirable due to the risk of warpages (non-flatness) or cracking of the sample during shrinking process. For situations where high firing shrinkage value may appear due to wide difference between unfired (dried) density and fired density, different or better shaping or compaction method must be introduced to reduce the shrinkage values.

Also, there are cases where expansion occurs during firing. This is due to some volume expansive reactions among the reactant phases (present in the composition) during firing, or due to phase transformations of the reactant or product phases. This expansion also needs to be taken care similar to that of shrinkage. Shrinkage is calculated as per the formula:

$$\text{Percent linear shrinkage} = (L_f - L_i) \cdot 100 / L_i$$

Where L_i is the initial length of the refractory before firing, and L_f is the final length of the refractory after firing.

TABLE 3.1

Specific Gravity Values of Some Common Refractory Materials

Name	Specific Gravity
Alumina (corundum), Al_2O_3	3.99
Silica (quartz), SiO_2	2.65
Silica (cristobalite), SiO_2	2.32
Silica (tridymite), SiO_2	2.28
Magnesia, MgO	3.58
Iron oxide, Fe_2O_3	5.24
Zirconia, ZrO_2	6.1
Chrome oxide, Cr_2O_3	5.22
Lime, CaO	3.34
Mullite, $3\,Al_2O_3\,2\,SiO_2$	3.16
Spinel, $MgO\,Al_2O_3$	3.58
Silicon carbide, SiC	3.21
Graphite	2.2

3.2.4 PERMEABILITY

The rate at which a fluid can pass through porous materials is termed as permeability. Refractories having a higher amount of connectivity of its pores are having higher permeability. Low-permeable refractories are essential for applications where they are in contact with gases and liquids.

3.3 MECHANICAL PROPERTIES

The mechanical properties of a refractory material describe how it will react to the mechanical forces that are prevailed in its application area. During use, refractories face varied mechanical conditions with varying degree of intensity, as they may be under compression, bending, shear, also sometimes partial tension and twisting. Moreover, the refractories are facing all these different conditions not separately, but all the mechanical actions are active simultaneously to varying extents at different portions in the same application environment. So a combined action of several types of mechanical forces is present to varying extent. The major mechanical properties that are commonly measured for refractories are detailed below.

3.3.1 COLD CRUSHING STRENGTH (CCS)

Cold crushing strength (CCS) indicates the compressive strength of a refractory material at ambient conditions or room temperature. Refractories have to withstand the structural load coming to them mainly from the load of the refractory lining or furnace and also the load of the charge or product material. This property reveals the idea of load bearing capacity of a refractory. CCS measures the bond ruptures strength of a material under compression, and it has an indirect relevance to refractory performance.

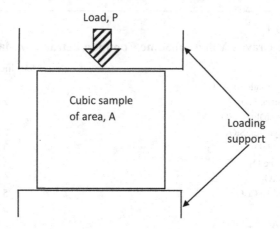

FIGURE 3.1   Schematic of the CCS measurement.

Though, many refractory technologists are not willing to give proper importance to this property, as dense refractories do not fail simply due to load at cold condition; but this easily performable test can be used to get some idea about the quality and predictable performance of the refractory. CCS can be used as one of the indicators of abrasion resistance. The higher the CCS of a material, the greater its bond rupturing strength, meaning the stronger will be the refractory and greater will be the resistance to abrasion, corrosion, etc. In general, higher sintering, higher densification results in higher CCS values. CCS is generally calculated on a cubic sample as per the formula:

$$CCS = P/A,$$

where P = compressive load at which the refractory sample disintegrates and A = the area on which load is applied. Figure 3.1 shows the schematic of the CCS measurement. Ceramics are brittle materials and have preexisting cracks due to their processing techniques. So, the bigger the size of the sample, the higher is the chance of having a larger sized crack, even from the same material processed under the same conditions. Hence, the obtained strength will be lower as the size of the sample increases. Hence, the sample size of any refractory needs to be specified for any mechanical testing.

3.3.2   Cold Modulus of Rupture (CMOR)

Modulus of rupture (MOR) or bending strength or flexural strength is an important mechanical property, especially for brittle materials, which measures the ability of the material to resist deformation under bending load. When the measurement is done at ambient conditions or room temperatures, then this is called as cold MOR. Here, a bar-shaped specimen with circular or rectangular cross-section is bent by applying the load from the top and two supports from the bottom side until fracture.

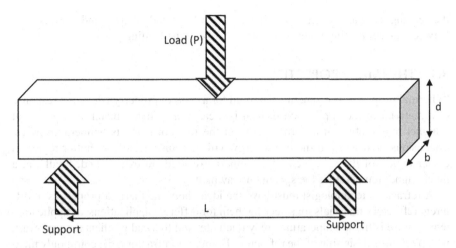

FIGURE 3.2   Schematic of the CMOR measurement for a rectangular bar sample.

Most commonly, the test method with two supporting points and one loading point is used for refractories called three-point bending method. In some special cases, mostly for critical structural applications, two separate loading points are used along with two supporting points called four points bending method. The flexural strength represents the highest stress that a material can withstand at its moment of rupture. Or in other words, it is the ultimate strength of a bar-shaped sample at its failure by flexure method, equal to the bending moment of rupture divided by the section modulus of the bar. A schematic diagram of CMOR measurement is shown in Figure 3.2. During bending under load, the sample extends to the lower surface and contracts in dimension on the top surface. Hence, the lower surface is under tension, and the upper surface in under compression. As the ceramic materials are weak in tension, the crack of failure under increasing load will start at the lower surface of the sample and then extends upwards till complete failure.

The important parameters for CMOR are loading rate (higher loading rate will result in higher strength values), the size of the sample (greater size will result in lower strength values), porosity (higher porosity will produce lower strength values), etc. Flexural strength (σ) is calculated by the formula:

$$\sigma = 3PL/(2bd^2) \text{ in 3-point test of rectangular specimen}$$

$$\sigma = PL/(\pi r^3) \text{ in 3-point test of round specimen}$$

$$\sigma = 3Pa/(bd^2) \text{ in 4-point test of rectangular specimen}$$

$$\sigma = 2Pa/(\pi r^3) \text{ in 4-point test of round specimen}$$

where P=bending load at breaking (rupturing) point, L=distance between the supporting points, b=breadth of the rectangular bar sample, d=depth (height) of the rectangular bar sample, r = radius of cylindrical rod sample, and a = (distance between the supporting points − distance between the loading points).

3.4   THERMAL PROPERTIES

"Thermal property" is a general term used for a class of property where the response of a material to the application of heat is measured. When a material is placed at a higher temperature, it absorbs energy in the form of heat, its temperature rises, dimensions increase, and the heat energy will be transported from hotter region to cooler region of the specimen (if a temperature gradient exists), and, finally, at a much higher temperature the specimen may melt.

A refractory technologist must have the idea about the thermal properties of different refractory materials for selecting them for different applications, as application temperature is high, temperatures may fluctuate, and thermal gradients may prevail, which requires designing of the refractory lining. For refractory, the commonly measured thermal properties are thermal expansion and thermal conductivity.

3.4.1   THERMAL EXPANSION

Whenever a material is heated, it absorbs heat energy, its energy level (potential energy) increases. This increased heat or thermal energy of the material results in higher vibrational movements of the atoms. That means thermal energy is converted to the vibrational energy (potential energy, PE) of atoms. This increase in vibrational energy increases the amplitude of atomic vibration of each atom of the material about their mean position. Figure 3.3 describes the potential energy plot against the interatomic distance of any material with varying temperature. The distance between two atoms varies during vibration, and the minimum and the maximum distances can be marked from the potential energy plot. Due to the asymmetric nature of this potential energy (PE) plot, the mean distance between the atoms increases from its stable most

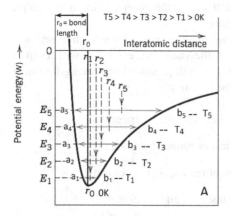 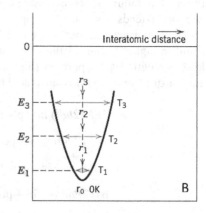

FIGURE 3.3 (a) PE plot and mean spacing between atoms for any material with increasing temperature; (b) PE plot of a strongly bonded material.

condition; thus, the mean distance between the atoms increases with temperature, and material expands on heating.

Say for a material the equilibrium distance of separation between its atoms (bond length) is r_0 at temperature $T=0$ K. As the temperature is increased (as marked by T_1, T_2, T_3, T_4, and T_5), the atoms absorb available thermal energy, the potential energy of the material is also increased (say corresponding PE levels are E_1, E_2, E_3, E_4, and E_5), and the vibration of atoms will also be increased to higher energy levels. Atoms vibrate between two extreme positions marked by the PE curve corresponding each energy level (Figure 3.3a). At temperature T_1, the atoms will vibrate between a_1 and b_1 positions, for T_2 temperature they will vibrate between a_2 and b_2 positions, and so on. Now, as the PE curve is asymmetric, the mean positions between the $a_1 - b_1, a_2 - b_2$, etc. (at higher energy levels), shift to the right side than that of equilibrium distance r_0. That is the separation distance between the atoms increases with increasing the temperature ($r_5 > r_4 > r_3 > r_2 > r_1$ when $T_5 > T_4 > T_3 > T_2 > T_1$). Thus, the dimension of the material increases with increasing temperature.

Again, the stronger the atomic bond between the atoms, the deeper and narrower will be the potential energy plot, and the lesser will be the asymmetry of the plot (Figure 3.3b). Hence, lesser will be the increase in mean distance between the atoms on increasing temperature and lesser will be the thermal expansion values. So for refractories and ceramics that are having stronger (ionic or covalent bonds) than that of metals (metallic bonds) have lower thermal expansion values than metals.

Thermal expansion is commonly expressed as,

$$\text{Percent linear thermal expansion} = (L_f - L_i) \times 100/L_i, \text{ and}$$

$$\text{Coefficient of linear thermal expansion} = (L_f - L_i)/(L_i \cdot \Delta T),$$

where L_i = initial length of the sample, L_f = final length after reaching the maximum temperature, and ΔT = temperature difference.

The magnitude of the coefficient of thermal expansion remains constant for any specific material in a specific temperature range but increases with rising temperature. At lower temperatures, this increase is due to the higher amplitude of vibration of atoms at a higher energy level (temperature). For higher temperatures, increased asymmetry in PE curve along with the formation of defects (Frenkel and Schottky) increase the value of coefficient of thermal expansion. Again for samples containing nonisometric crystals, thermal expansion values vary with the crystallographic axis. The closed packed direction (higher bond strength) has a lower value of thermal expansion, and this variation in the expansion values reduces with increasing temperature. The coefficient of thermal expansion values of some common refractory materials is given in Table 3.2. Also, linear thermal expansion of different refractory materials is shown in Figure 3.4.

3.4.2 THERMAL CONDUCTIVITY

Thermal conductivity is an intrinsic property of a material and related to the transfer of heat. Heat flows from a higher side to a lower side, and the material that carries

TABLE 3.2
Coefficient of Linear Thermal Expansion Values of Some Common Refractories

Name	Coefficient of Linear Thermal Expansion (α), $°C^{-1} \times 10^{-6}$
Alumina (corundum), Al_2O_3	8.8
Fused Silica	0.4
Magnesia, MgO	13.5
Zirconia, ZrO_2 (stabilized)	11
Mullite, $3\,Al_2O_3\,2\,SiO_2$	5.3
Spinel, MgO Al_2O_3	7.6
Silicon carbide, SiC	5.12

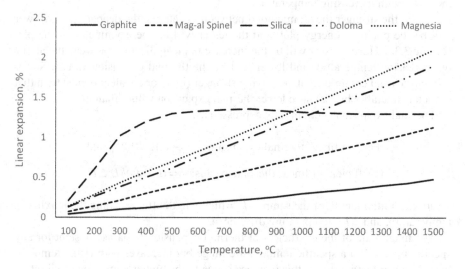

FIGURE 3.4 Linear thermal expansion of different refractory materials against temperature.

the heat is called conductor. Thermal conductivity is the measure of the ability of a material to carry the heat from its hotter side to cooler side. Heat conduction takes place when a temperature gradient exists between two opposite surfaces of a solid (or stationary fluid) medium. Conductive heat flows from a high temperature to a low-temperature region, as the higher temperature is associate with higher molecular energy or greater atomic movement. Energy is transferred from the higher energetic atoms to the less energetic ones when neighboring molecules collide or come into contact with each other. Thermal conductivity is described as the amount of heat energy transferred per unit time per unit surface area separated by a unit distance having a unit difference of temperature.

Thermal conductivity [K, unit W/(m.K)] describes the transport of heat energy (Joule) passing through a body per unit time (second) per unit area (square meter)

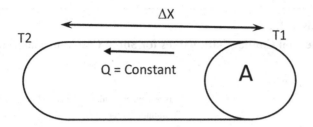

FIGURE 3.5 Sketch of heat flow (thermal conduction) through a material.

separated by unit distance (meter) under a unit temperature gradient (Kelvin), as shown in Figure 3.5. It is expressed as the amount of heat (Q) transported between two surfaces (flowing perpendicularly at a steady rate) per unit of time (t) [that is dQ/dt or heat flow] through an unit area (A) of cross section [so dQ/A.dt, that is heat flux] at an unit temperature gradient separated at a unit distance [$\Delta T/\Delta x$, where ΔT is the temperature difference between two surfaces (T_{1-} T_2), and Δx is the distance between two surfaces]. So thermal conductivity,

$$K = (dQ/dt)/[A \cdot (\Delta T/\Delta x)]$$

As per thermal conductivity values, materials are classified into two major groups. Materials having high thermal conductivity are called conductors, used mainly to conduct/transfer heat from one side to the other and also act as a heat sink. The other types are having very low conductivity values called insulators, which are used for thermal protection, to prevent any heat loss from any system (the best examples are refractories). Approximate values of thermal conductivity of some common materials are given in Table 3.3. Also, variation of thermal conductivity of different refractory materials against temperature is shown in Figure 3.6.

3.5 THERMOMECHANICAL PROPERTIES

The most significant properties of refractories are those that allow them to work in the application environments at elevated temperatures. Among them the ones withstanding mechanical load at high temperature is most important. Refractories must have sufficient strength to withstand the load of the furnace structure and charge/product material, and also need to bear the mechanical action of the charge and process materials at the processing conditions. Hence, evaluation of mechanical properties at the application temperatures is a critical parameter for selecting the refractory.

Refractories are mostly prepared from natural materials, and the presence of impurities, even in a minute amount, is very common. Though the raw materials for a specific refractory are very pure, a minute amount of impurity is effective in forming the little amount of low melting phase in the refractory. This liquid phase is sufficient to cause deformation of shape and degradation in strength of the refractory at the elevated temperatures. Both of these will result in sagging or collapse of the refractory structure. Hence, measurement of strength at high temperatures gives

TABLE 3.3
Thermal Conductivity Values for Some Common Materials

Thermal conductivity at RT, W/m.K	
	Graphene
1000	
	Silver, copper, gold, silicon carbide
	Aluminum, graphite
100	Silicon
	Metals, iron,
	Steel, carbon bricks
10	Silicon nitride
	Aluminosilicates
	Refractories
1.0	Ice, glass, fire clay, concrete,
	Water
	Polymer, coal, brick, epoxy
0.1	Polyethylene, building boards, oils, Wood
	Fiber boards, insulations
	Air, polystyrene, organic foam, paper, cotton
0.01	Krypton, Freon (gaseous)
0.001	Vacuum insulation

the idea of the safe highest application temperature of a refractory. There are two common techniques for measuring strength at high temperatures. These are the hot modulus of rupture and compressive creep.

3.5.1 HOT MODULUS OF RUPTURE (HMOR)

This is nothing but the measurement of modulus of rupture, as described in cold MOR (Chapter 3.3.2), in hot condition. The whole testing is done in a furnace at the desired test temperature. Three-point bending test is done on rectangular bar samples for HMOR. This measurement is especially important, as it also incorporates the tensile condition of the refractory sample along with the compressive one, which is not included in other hot-strength measurement techniques for refractories. Specially designed furnaces named HMOR furnace are used for this testing having sample support and loading rods inside the furnace made of high-temperature ceramic materials. The formula and calculation used for HMOR is exactly the same as that of the cold MOR.

3.5.2 CREEP

The common feature of the application conditions of different refractories is high temperature and load. And this condition continues for a long time, and in some cases, it is years together. Hence occurrence of any deformation under heat and load for prolonged period is important for understanding the structural integrity of the refractory

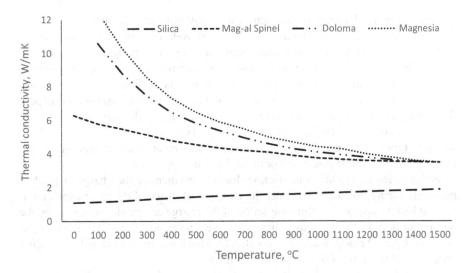

FIGURE 3.6 Variation of thermal conductivity of different refractory oxides against temperature.

lining. Creep is the measurement of the deformation of any material against time under a specific load and temperature conditions. Hence, for refractories creep is one of the ideal characterization technique to evaluate its performance. The creep measurement is little different for refractories than to that of the metals, as for refractories it is conventionally compressive creep, but for metals, it is generally in tensile mode.

When a material is placed under a load of fixed stress value at a temperature above $0.4\ T_m$ (where T_m is the melting point of the material in Kelvin scale), the strain of the material continues to increase with increasing time at a rate depending on the inherent characteristics of the material. This slow and continuous deformation of a material against time under constant load and heat is defined as a creep. Creep is an anelastic property of a material and sometimes also expressed as viscoplasticity of a material. This is because when a material deforms under heat and load, it behaves like a plastic material (viscous flow), and the deformation is permanent in nature (not an elastic one). This deformation is due to the presence of liquid phase in the material at high temperature under load resulting in sliding of grains and deformation. Again deformation occurs due to the reduction in viscosity of the liquid under load at high temperatures. A refractory material is better creep resistant even on higher amount of liquid phase formation if the liquid has a higher viscosity (resulting in lower flowability and deformation). So the amount and viscosity of the liquid present (that depends on the type and amount of the impurity phases) control the deformation and creep behavior of the refractory materials to a great extent.

3.6 ABRASION PROPERTIES

Abrasion is the damage or destruction of a material due to friction against another material (surface) during use. It is related to the interactions between surfaces, and caused due to the deformation and removal of material from the surface as a result of the mechanical action of the abrading surfaces. Relative motion between the two surfaces and initial mechanical contact between them are important. Abrasion wear can also be defined as a process where interaction between two surfaces or bounding faces of solids within the working environment results in dimensional loss of one solid, with or without any actual decoupling and loss of material. Abrasion is dependent on the conditions of the working environment, such as the direction of sliding; nature of load (reciprocating, rolling, and impact); speed; temperature; etc. Also, different types of counter-bodies such as solid, liquid, or gas, and type of contact ranging between single phase and or multiphase are important.

Refractories face continuous friction due to movement of the charge or product materials during use, and also under the mechanical thrust of the processing conditions at high temperatures. Rubbing action of the charge and product particles on the porous surfaces of the refractory material causes the removal of material and wear. Descending solid charge materials in a blast furnace is a common example for sliding abrasion and wear of refractory present on the wall.

Also, refractories face impact wear, which is, in reality, a short sliding motion where two surfaces interact at an exceptionally short time interval but with a higher mechanical thrust. Wear increases with sliding distance between the surfaces, load, and is inversely proportional to the strength, densification, and hardness of the refractory. Wear is defined as the loss of substance from surfaces in contact with relative motion. When two different surfaces, one hard and another soft, are in contact and relative motion, the softer asperities undergo fracture or deformation and wear. Dense, high strength, fine grained, high hardness refractories show better wear resistances. Loss of refractory material due to mechanical action (wear), measured in weight, and expressed as volume loss is defined as the abrasion loss of refractory.

3.7 CORROSION PROPERTIES

Corrosion behavior of any refractory is important for the performance and life in any application area. Corrosion behavior is the chemical property of any refractory, dependent on multiple factors and cannot be determined just by viewing or touching the substance. As it affects the internal structure of any material (unlike the physical properties) and may affect all other properties, detail understanding and evaluation of corrosion behavior of refractory are required.

Refractories are used for any high-temperature processing where they face different chemical environments and are in contact with chemically active solids, liquids, or gases. Hence, for refractories, resistance to these chemicals are very important for their structural integrity, performance, and also important for the refractory/furnace structure. Refractories need to have chemical corrosion resistances that are prevailed in any specific application. A chemical reaction with the solid/liquid/gaseous environments may result in new phases in the refractory, volumetric instability due to change in volume, the formation of the liquid phase, deformation, and obviously degradation of properties. A direct reaction between solid materials with refractories are less probable due to less number of contact points, but corrosion in contact with gases and liquids is prominent and important for the performance and life of the refractory. There are different techniques available for characterizing the corrosion resistance of a refractory material. As per the application environments, refractories that are used for the stack area of a blast furnace (gaseous attack from carbon monoxide), and refractories that are in contact with any slag or glass (liquid phase) are most important for the evaluation of chemical resistances.

3.7.1 CARBON MONOXIDE (CO) DISINTEGRATION

The best example of corrosion of refractory by the gaseous material is the upper stack area of blast furnace where the refractories are attacked by carbon monoxide (CO) gas prevailed in that region (CO is generated from the carbo-thermal reduction reaction of iron-making process in a blast furnace.). The degradation in quality and disintegration of the refractory by CO is mainly dependent on the amount of ferric oxide present in the refractory.

The maximum temperature in the upper stack area of blast furnace reaches up to around 600°C, and fire clay-based refractories are most suitable for such low-temperature applications from the property requirements and economic points of view. The natural sources of fireclay are contaminated with good amount of iron oxide, and this iron oxide reacts with CO present in the environment at the temperature range of 450°C–550°C to form iron carbide (cementite) phase. This reaction is associated with volumetric expansion that causes cracking and disintegration of the refractory.

$$6\,Fe_2O_3 + 4\,CO = 4\,Fe_3C + 11\,O_2$$

Even though there is no disintegration of the refractory, there may be decoloration or carbon deposition in it, which drastically deteriorates the properties. So fire-clay refractories with a limited amount of iron oxide are to be selected for such applications.

3.7.2 SLAG OR GLASS CORROSION

In many of the high-temperature applications, refractories are in contact with liquid phases and in mostly with corrosive liquids. Very common examples are slag (which is mainly basic in nature for iron and steel industries), glass (which is acidic in nature), etc. These chemically active liquids at high temperatures react with the refractory materials and cause corrosion. These liquids can also penetrate within the refractory through the pores and corrode the structure, resulting in drastic deterioration of the properties. Hence, it is important for any refractory manufacturer and user to know how much the corrosive liquid can penetrate or react, and corrode and degrade the refractory.

Any liquid can penetrate a refractory if the liquid is highly fluid, and there are openings (like pore or cracks) on the refractory surfaces. Hence, penetration of liquid slag or glass is dependent on the amount and size of surface pores (apparent porosity) and cracks. The corrosive liquid can only penetrate through a pore or crack if the size of the pore or crack is larger than that of the minimum droplet size of the slag or glass at that temperature (consideration of surface energies). Higher the temperature, lower is the viscosity of the liquid, and lower will be the minimum size of the droplet, so higher is the (chance of) penetration. Slag- or glass-penetrated refractory portion will have a different character than that of the rest unaffected refractory, and causes cracking, disintegration and degradation of the whole refractory structure.

Again refractories, as a whole or any component of it, may react with the corrosive liquid (slag or glass) and may get dissolved or washed (eaten) away. Impurities present in the refractories are prone to be attacked by the corrosive liquids and degrades the quality of the refractory faster. Corrosion reactions proceed through the matrix phase or grain boundaries of the refractory microstructure, where the concentration of the impurities is high, though the main granular phase may remain unaffected and unaltered. Microstructural studies can confirm exactly which component of corrosive liquid is more penetrating or more corrosive to the refractory, or which types of impurity phases of the refractory are affected most, and what kind of reactions are occurring within the refractory microstructure due to the presence of the corrosive liquid or its components. For the case of alumina refractory being corroded by steel-making slag, if lime-bearing aluminosilicates are found in the corroded refractory portion, then lime and silica are the dreaded components of slag that cause the degradation most. Again, if the formation of calcium hexa aluminates is found in microstructure/phase analysis study, it is beneficial to restrict the progress of the penetrating slag and its components due to the volume expansion during the formation, filling the pores and voids, and limiting the ingress of slag in the refractory.

3.8 MICROSTRUCTURE

Microstructure is the microscopic description of the individual constituents of a material. In other words, it is the very small-scaled structure, as observed on a prepared surface of any material under a microscope. It describes the structural characteristics at a very minute level and is defined by the type, proportion, and composition of the phases present and by their form, size, and distribution. Microstructural characterization of the refractories is an important parameter, as it determines the development of properties, like, porosity (density), strength, toughness, hardness, corrosion resistance, thermal properties, wear resistance, etc. In turn, microstructure governs the application of refractories in industrial practice. In microstructure, we study the arrangement of the phases present in a refractory at that temperature. This arrangement of phases again depends on various factors like, raw materials used, fabrication techniques and temperature of processing, phase–equilibrium relations, sintering and grain growth, different phases developed, kinetics of phase changes, final phases present, and their distribution, etc. Thus microstructural development of a refractory is a complex phenomenon, and its evaluation is also very important to understand a refractory and its properties. The microstructure of refractory materials is generally evaluated by determining,

1. The number of different phases present, including porosity,

2. Characteristics of each phase, like, shape, size, orientation, and other features
3. Relative amount of each phase present
4. Features of the interphase and intraphase boundaries.

Common features of ceramic microstructures are (i) predominating crystal phase, (ii) secondary crystalline phases, (iii) matrix phase or glassy phases, (iv) pore, (v) grain boundary, etc. In general, microstructure of refractories is coarse in nature when compared with common ceramic products like whitewares, porcelain, and claywares. Refractories are mostly made up of different sized grains and fines; the filler grains are generally in the size of micrometers to millimeters (a representative figure for refractory microstructure is shown in Figure 3.7) in comparison to micrometer-sized particles in clay-based ceramic wares. Moreover, the refractories are mostly porous, having good amount of air spaces of varying shape, size, and amount in between the grains. Final microstructure of a refractory product depends on many parameters, like

1. raw materials used (particle size, impurity) and their amount
2. fabrication technique (mixing sequence and proportion of raw materials, shaping, and firing conditions)
3. phase–equilibrium relations among the constituents
4. kinetics of phase change
5. grain growth
6. heat treatment (temperature, time, atmosphere)

The microstructure of refractories can be designed by controlling different aspects like, optimizing the granular composition, modification of processing parameters, addition of additives, etc., and less porous, compact microstructure can be designed and obtained with optimized and improved properties.

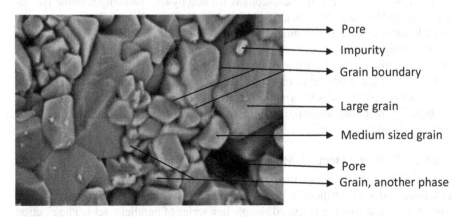

FIGURE 3.7 Representative photograph for refractory microstructure (scanning electron micrograph).

3.8.1 Techniques for Evaluation of Microstructure

Evaluation of microstructure for refractories is done by the common techniques that are used for ceramics and materials science. There are three major categories of microscopy study for refractories that are described below.

3.8.1.1 Optical Microscopy

In this technique, microstructural study of the refractories is done by using visible light to provide a magnified image of the micro- and macrostructure using optical lens. Here, the illuminating source is visible light with wavelengths from 450 to 600 nm. The two most widely used methods for optical microscopy are,

    a. Observations of thin sections under transmitted light, and
    b. Observations of polished sections with reflected light.

The thin section of the refractory sample should be between 0.015 and 0.030 mm thickness, so that ordinary light can pass through it. The sample is prepared by cutting a thin slice, polishing one side of it, cementing this side to a transparent microscopic (glass) slide, and grinding and polishing the other side to obtain a section of the required uniform thickness. The advantage of this method is that the optical properties of each phase present in the sample can be studied, and the phases can be identified.

The polished sections are prepared by grinding and polishing one surface of the cut refractory sample (mounting on the cut specimen in Bakelite or Lucite plastic) using a series of abrasive powders suspended in water on cloth wheels and abrasive pastes. The polished surfaces are observed under reflected light in a microscope to distinguish the differences in relief or reflectivity among the phases present. Etching of the sample (chemical or thermal) reduces the impurity phases from the sample, and improves the features of the microstructure and clarity of differences among the different phases.

The resolution in optical microscopy is limited by the wavelength of the incoming light, and because the optical light has a higher wavelength, optical microscopes has a lower resolution (~200 nm). Moreover, as the focal length of the optical lens is small, and the physical size of the microscope is limited, the magnifying capacity for optical microscopy is also limited, up to a magnification of about 2000 times. Use of polarized light in optical microscopy helps in easy phase identification. Polarized light microscopy provides information on absorption color and optical path boundaries between the phases of differing refractive indices, and can also identify isotropic and anisotropic substances.

3.8.1.2 Scanning Electron Microscopy (SEM)

In this evaluation technique, the polished/fractured surface of refractory specimen is bombarded with artificially generated and focused beam of high-energy electrons that scan across the prepared sample in a series of parallel tracks. The electron interact with the surface atoms of the sample, collect information from the surface, and generate different signals that can be detected and displayed on the screen of a cathode ray tube/electronic screen (display unit). The illuminating source for SEM is electron beam with energy level between 1 and 30 keV; it has wavelength in angstrom units (AU); and the resolution of SEM is much higher than optical microscopy and in the range of 4–1.5 nm. Also, the magnification is the ratio of the image screen size of the display unit to the area of the sample on which the electron beam is falling, resulting in a magnification even up to 1,50,000 times.

Refractories (ceramics) are generally nonconductors, so the electrons after bombardment get accumulated on the surface of the sample, results in electrostatic charging, and repel the incoming electron beam. This will lead to charging of the surface and cause obstruction in getting the sample surface features. To avoid such disturbances, the sample surface is coated with a thin layer of highly conducting material (carbon, gold, platinum, etc.) with a conducting path on the side surface to make a passage for the accumulated electrons to pass through to the metallic sample support piece at bottom. Again, as SEM is using high-energy electron beam, having very small wave length, polished surfaces are not necessarily required to generate the image, and fracture surfaces can be used. Also, small wave length is very useful for getting the minute surface features.

In SEM, two types of electrons are primarily used for evaluation of microstructure, namely, secondary electrons (SE) and backscattered electrons (BSE). Secondary electrons originate from the atoms of the sample surface due to inelastic interactions between the electron beam and the sample. Whereas the backscattered electrons are reflected back after elastic interactions between the beam and the sample. BSEs come from deeper regions of the sample, while SEs originate from surface regions. Hence, these two types of electrons reveal different types of information. Images formed by SEs provide details of the topography, that is, surface features and information. Again, BSE images show high sensitivity to atomic number; the higher the atomic number, the brighter the surface appears in the image. Moreover, BSE images reveal about crystallography, topography, and the magnetic field of the sample.

In SEM, several signals are also generated due to interaction between the focused beam of high-energy electrons and the electrons of the sample surface. These signals reveal different information about the sample, including external morphology (texture), chemical composition, and crystalline structure and orientation of materials making up the sample. Thus SEM is also capable of performing various analyses of selected point locations on the sample. It is useful for determining chemical compositions in qualitatively or semiquantitatively [using EDS (energy dispersive X-ray spectroscopy)], crystalline structure, and crystal orientations [using EBSD (electron back scatter diffraction)].

Microscopic study through SEM is further improved by using field emission scanning electron microscope (FESEM) to obtain further detailed information on the microstructure. It is a modified SEM that provides higher resolution, increased magnification, and a much greater energy range. The main difference between a FESEM and a SEM lies in the electron generation system; a field emission gun is used in FESEM that provides extremely focused high- and low-energy electron beams, which greatly improves spatial resolution and enables work to be carried out at very low potentials (0.02–5 kV). This also reduces the charging effect on nonconductive specimens and minimizes the chance of damage to the sensitive samples by electron beam. Also, FESEM uses in-lens detectors, which are optimized to work at high resolution and very low acceleration potential, thus providing better microstructural features. In comparison to a typical SEM, the FESEM has much brighter electron source and smaller beam size, thus increasing the useful magnification for observation even up to 500,000 times.

3.8.1.3 Transmission Electron Microscopy (TEM)

The TEM is a very powerful tool for material science; however, it is not commonly used for the study of refractories. For very specific requirement or understanding in atomistic level, this tool is used in refractories. In this technique, a high-energy beam of electrons is passed through a very thin specimen, and the transmitted beam coming out from the sample is analyzed for structural information. The interactions between the passing electrons and the atoms can be used to observe features such as the crystal structure and features in the structure like dislocations and grain boundaries. TEM can be used to study the growth of layers, their composition, and defects associated. TEM provides the selected area electron diffraction (SAD) patterns that helps to identify and characterize individual phases present in the structure.

The TEM operates on the same basic principle as that of the optical microscopy under transmitted light but uses electrons instead of light. The illuminating source for TEM is electron beam, but with higher energy level compared to SEM, in the range of 100–300 keV. As the wavelength of high-energy electrons is much smaller, the resolution attainable for TEM is in the range of 0.2–0.07 nm, far better than SEM and many orders of magnitude better than that of optical microscope.

TEM study provides the finest details of internal structure; in some cases as small as individual atoms. But the sample thickness required for TEM is <1 µ, which is more than an order of magnitude lesser than that required in optical microscopes. This is because of the lesser transmittance value for electrons than that for photons. Such a thin thickness is obtained by chemical thinning or thinning by argon ion bombardment after polishing and grinding of the sample. However, the TEM equipment and the characterization is cost intensive for industrial and bulk samples, like refractories; characterization through TEM is not common.

3.8.2 Components in Refractory Microstructure

Refractories are polycrystalline materials and generally a multiphase system. Porosity is an obvious component in the system, as refractories are made up from granular mixes, and the gap between the grains remain within the sintered body resulting in porosity in the system. Refractories are generally a multiphase system, containing different phases other than porosity. Even a very high, pure, monocomponent system (say high alumina refractory) contains some amount of impurity phase in the system (coming from the raw materials) and may form a number of secondary impurity phases that remain within the microstructure. Also, some refractories may contain little amount of glassy phase in the final microstructure, depending on the impurities present and reactions between the constituting components with the impurities forming liquid phase. Brief description on the components of refractory microstructure is given below.

3.8.2.1 Porosity

Porosity is the void portion present in a solid mass, quantified as the volume fraction of pores (voids) present per unit volume of the total mass (solid + pores). Porosity is characterized by the size, shape, and distribution of the pores. In refractories, porosity can vary from 0% to > 90% of the total volume of the material, but the extreme cases are rare in applications. Presence of porosity directly affects all the different types of properties, like density, strength, conductivity, thermal expansion, corrosion, wear, etc. Commonly porosities are observed as black (dark) phases under microscope, and are mostly seen at the junctions of the grains or at the grain boundaries. Size of the pores and their features, like isolated, spherical, channel, intergranular, intragranular, etc., helps to infer about the processing of the refractories and provides information on the probable characteristics.

3.8.2.2 Single-Component Refractory

In the case of single-component system, the refractory microstructure ideally contains the stable phase of the component along with porosity. To understand the microstructure, the size, shape, and distribution of the granular phase (observed as grains) and the porosity need to be evaluated. Grains may be columnar, prismatic, cubic, spheroidal, or acicular in shape, and different shapes may result in different properties for the same material. Again, presence of impurity (or additive) phase may result in minute amount of secondary phases (may be glassy phase too) and affect the properties of the refractories.

3.8.2.3 Multicomponent Refractory

In multicomponent refractory systems, the relationships among the phases, relative amount of the phases, their distribution, and orientation of each phases are important and need to be considered for evaluation. The most common feature observed in multicomponent system is that one or more phases are dispersed in a continuous matrix. Increase in number of components in the system makes the evaluation and understanding of microstructure complicated. Each phase in microstructure has distinct effect on the final properties. Again, presence of impurities or additives may also affect differently for different components and affect the properties differently. Hence, detailed study of microstructure is required for a multicomponent system to characterize and understand the refractory.

3.9 REFRACTORY-SPECIFIC PROPERTIES

Till now the properties described are of generalizes nature and common to other materials also, and they are studied in certain other fields too. But there are some properties commonly used and measured only for the refractory materials. These refractory-specific properties are not required, useful, and measured for any other types of materials. These properties, which are described below, are very common and conventional to refractories, and are commonly used in any specification of the refractories.

TABLE 3.4
Melting Point Values of Some Pure Refractory Materials

Name	Melting Point (°C)
Alumina (corundum), Al_2O_3	2050
Silica, SiO_2	1713
Magnesia, MgO	2825
Zirconia, ZrO_2	2550
Chrome oxide, Cr_2O_3	2427
Lime, CaO	2570
Mullite, $3 Al_2O_3 2 SiO_2$	1810
Spinel, $MgO Al_2O_3$	2135

3.9.1 REFRACTORINESS OR PYROMETRIC CONE EQUIVALENCE

It is the inherent property of any material by which it can withstand the heat effect (high temperature) without appreciable deformation or softening at no external load condition. Refractoriness is also called softening temperature or pyrometric cone equivalent (PCE) temperature and is measured as a specific temperature that indicates the starting of fusability of any refractory material without any external load. In other words, it is the temperature when a material starts to soften under the action of heat under only its own weight. For a very pure material, the refractoriness matches (very close) with the fusion (melting) temperature. Hence, this property may also come under the broad category of thermal properties of refractories. Table 3.4 shows the melting (fusion) temperature of some commonly used refractory materials.

Refractories are made of natural materials that are commonly associated with impurities. Even in the case of synthetic materials, purity level is significantly high, but the material is not free from impurities. So refractories are always associated with some amount of impurities. Again, the presence of an impurity or a secondary material generally lowers the liquid formation temperature, as commonly found in phase diagrams. Commonly, the refractories available commercially (having impurities in them) do not have any sharp melting temperature and start to soften at a lower temperature (melt progressively over a range of temperature) than that of the pure materials' melting point. Effect of the impurity/secondary material on the softening of refractory materials depends on the following factors,

1. Chemical nature of impurity phase (drastic reduction in softening point occurs when the impurities are of opposite chemical nature to that of the main constituent, say a basic impurity in an acidic refractory and vice versa),
2. Amount of the impurities present,
3. Fusion point of the lowest fusing constituents present in the system, and
4. The capacity of the lowest fusing constituent to react and dissolve the higher fusing components.

Refractoriness is measured by comparing the softening character of an unknown material against some known materials, making the samples as a pyramid-shaped cones. These pyramid-shaped cones are termed as pyrometric cones. As the test method finds the equivalence in softening behavior of the test sample (unknown) against the standard sample (known) in the shape of cone, this property is termed as pyrometric cone equivalent (PCE). This pyrometric cones were first developed by Dr Herman August Seger and accordingly the standard cones are named as Segar cones. Also, there are Orton cones that are globally accepted and used. Each standard cone has a number that corresponds to a specific temperature where that specific numbered cone will soften. This is done by very careful and precise selection of high, pure raw materials and mixing them in such a proportion that the particular numbered cone will soften on that specific temperature only. The softening temperature of a specific cone number may vary for different standard cone systems (like Seger, Orton, etc.), as they are manufactured with different compositions and from different sources of raw materials.

The pyrometric cones have a triangular base, and a defined dimension and size. On heating, the cones that are placed at a little inclination (not vertical) will bend due to softening at high temperatures. This bending and lowering of the tip will continue toward the inclined direction with increasing temperature. Now, the tip of a specific numbered cone will touch the base at a specific temperature. Every cone number of any standard cone system has a fixed tip-touching temperature (at which the tip touches the base). Any test (unknown) sample's cone when matches with a standard cone in tip-touching temperature, the unknown sample is referred to have the refractoriness equivalent to that standard cone number (or its temperature). When the refractoriness (tip-touching temperature) of the unknown sample falls in-between two standard cone numbers, it is expressed as in-between those two standard cone numbers (associated temperatures).

Cone manufacturers strictly follow the raw materials' specification and specific manufacturing process to control the variability (within batches and between batches) and to ensure that cones of a given grade remain consistent in their properties over long periods. Even though cones from different manufacturers can have relatively similar numbering systems, they are not identical in their softening characteristics. If a change in the cone system is done from one manufacturer to another, the differences in tip-touching temperatures must be considered. Table 3.5 shows the cone numbers and their corresponding temperatures of commonly used standard cone systems.

Though PCE gives an idea about the softening temperature related to the maximum allowable temperature limit of a refractory (without any external load) but for practical applications, this temperature has little relevance. There is no such application where a refractory is used without any external load. Hence this measurement, though gives an idea for the inherent characteristic of a particular refractory material, is not very useful for practical applications.

3.9.2 Refractoriness under Load (RUL)

Refractoriness under load (RUL) is an important property of a refractory from the practical application point of view. This property measures the refractories

TABLE 3.5
End Point of PCE Measurement for Different Cone Systems
Used in Refractories

Cone No.	End Point (°C)			
	Orton (US)	Segar (Germany)	British	French
12	1337	1375	–	–
13	1349	1395	–	–
14	1398	1410	–	–
15	1430	1440	–	–
16	1491	1470	–	–
17	1512	1500	–	–
18	1522	1520	–	–
19	1541	1540	–	–
20	1564	1560	1530	1530
23	1605	–	–	1580
26	1621	1580	1580	1595
27	1640	1600	1610	1605
27½	–	1620	–	–
28	1646	1640	1630	1615
29	1659	1660	1650	1640
30	1665	1680	1670	1650
31	1683	1700	1690	1680
31½	1699	–	–	–
32	1717	1710	1710	1700
32½	1724	1720	–	–
33	1743	1730	1730	1745
33½	–	1740	–	–
34	1763	1760	1750	1760
35	1785	1780	1770	1785
36	1804	1800	1790	1810
37	1820	1830	1825	1820
38	1835	1860	1850	1835
39	1865	1880	–	–
40	1885	1900	–	–
41	1970	1940	–	–
42	2015	1980	–	–

of a refractory under a specific load. In this characterization technique, initiation (measured as temperature) of deformation or softening of a refractory composition is evaluated under a constant load against increasing temperature. As both temperature and load are involved in RUL, it may come under the broad category of thermo-mechanical properties. RUL indicates the safe temperature of use for a refractory under the combined effect of heat and load. The specific load maintained for RUL is 0.2 MPa.

In practical application, refractories are always under some extent of load, from the dead weight of the furnace and refractory lining, from the charge/product materials, and from the fuel, air, flue gas pressure, etc. Moreover, this load is unevenly distributed throughout the refractory lining and may change with time; sometimes cyclic loading is also there (say for rotary kilns, steel ladles, etc.). Aggregates and grains within the refractory structure will slide over one another under the application of external load whenever a minute amount of liquid phase is present in-between them (low melting compounds may form due to the presence of impurities) at high temperatures. This will cause a deformation of the refractory shape and will limit the application of the refractory at further high temperatures. Again, the viscosity of the liquid phase also plays a vital role on sliding of grains and deformation of refractory. The very minimum amount of impurity and high viscosity of the fused mass/liquid phase will result in minimum deformation and excellent RUL values in refractory. The measurement of deformation is calculated from the dimensional change of the refractory during heating under load. Two different temperatures, temperature of appearance (T_a) and temperature of the end (T_e), are specified as the initiation of deformation and end of deformation of the refractory. T_a has greater importance, as this can be taken as the highest safe temperature of application for any refractory. Refractoriness and RUL (T_a) values of some common refractories are shown in Table 3.6.

In general, refractories have a lower RUL value compared to its refractoriness (PCE value). Deformation due to sliding of grains under load at high temperatures on the formation of a liquid phase in the refractory matrix is responsible for such a wide difference. Refractories deform at a much faster rate, in presence of a liquid phase even present in a very minute amount when they are under mechanical load. Hence, RUL is a much better parameter to judge a refractory than refractoriness. But, still the measurement technique of RUL is far away from actual service conditions. In application, the load on refractories is not constant for its life, may also vary from one portion to another, and may also be cyclic in nature (say in rotary kilns). Also, the whole refractory may be not in a temperature equilibrium condition; only one surface may be getting heated; and the major portion of the load may be carried out by the relatively cooler portion of the refractory. So RUL, though better than refractoriness, gives only an indication of the temperature at which the bricks will start to deform in service conditions with similar kind of fixed load. It only gives a comparative index of quality among different refractories. Also, RUL does not consider any time factor, which is equally important at that high temperature. Use of refractory for prolong time is the general requirement, and deformation starts at a much lower temperature on prolonged use due to fatigue. So judging of the actual performance of a refractory is difficult to understand from the RUL value. In this regard, evaluation of creep (discussed in Section 3.5.2) is better as it incorporates time factor as a function for the measurement of deformation at high temperature under load.

3.9.3 THERMAL SHOCK RESISTANCE

This property can also be called as thermal spalling resistance or thermal fatigue resistance. Degradation in the quality of a refractory due to repeated (cyclic) change in thermal state is measured by this property. During use refractories often face

TABLE 3.6

Refractoriness and RUL (Ta) Values of Some Common Refractories

Refractory/Major Constituent	Purity (%)	Refractoriness, °C	RUL, (Ta), °C
Silica	SiO_2_94	1680	1650
	SiO_2_97	1710	1680
Fire clay	Al_2O_3_30	1680	1350
	Al_2O_3_40	1740	1450
	Al_2O_3_45	1760	1480
High alumina	Al_2O_3_50	1760	1500
	Al_2O_3_60	1785	1420
	Al_2O_3_70	1804	1550
	Al_2O_3_80	1820	1550
	Al_2O_3_85	1820	1580
	Al_2O_3_90	>1850	1700
	Al_2O_3_95	>1850	1720
MgO	90	>1850	1600
	95	>1850	1650
	97	>1850	1680
$MgO-Cr_2O_3$	MgO−60 (minimum) Cr₂O₃−15 (minimum)	>1850	1700
	MgO−50 (minimum) Cr₂O₃− 6 (minimum)	>1850	1650
$MgO-Cr_2O_3$ (direct bonded)	MgO−60 (minimum) Cr₂O₃−20 (minimum)	>1850	1720
	MgO−72 (minimum) Cr₂O₃−11 (minimum)	>1850	1700
	MgO−55 (minimum) Cr₂O₃−24 (minimum)	>1850	1750

heating and cooling cycles, rapid change in temperature that causes uneven expansion and contraction of the refractory, resulting in huge thermal stress and associated strain, and finally cracking, fracturing, and breaking of the refractory. This strain may be more critical if the refractory composition is not uniform, as it may cause nonuniformity in the thermal expansion character. This nonuniform expansion behavior may result in cracks and failure of the refractory.

Thermal spalling is dependent on the environmental conditions and process parameters where thermal cycles are occurring, like the temperature difference between hot and cool conditions, heating and cooling rate of thermal cycles, the medium of thermal cycle conditions (like, water, air), etc. Thermal spalling is also dependent on the following refractory properties.

a. Thermal expansion: Higher the thermal expansion characteristics, higher will be the dimensional change of the refractory with change in temperature.

Now, if the temperature drop is very fast, the refractory will have nearly no time to adjust the dimensional change and will have a tremendous strain within it. This strain will keep on multiplying as the number of the thermal cycles will progress and will finally result in cracking/breaking/bursting of refractory. Hence, to have a better thermal shock resistance, refractory should have low thermal expansion characteristics.

b. Thermal conductivity: During thermal cycles, the surface of the refractory will reach the outside environment temperature instantly, but the interior will change slowly and will remain in the previous conditions for prolonged period. This difference in thermal state between the interior and the exterior of a refractory will cause strain and may lead to failure. Now, if the thermal conductivity of the refractory is high, the interior of the refractory will also reach the ambient thermal conditions very fast, resulting in a very minimum strain, reduced cracking, and reduced chances of failure. Hence, for better thermal shock resistance, refractory must have high thermal conductivity values.

c. Strength: If the refractory has a higher mechanical strength characteristic, crack that may be generated by the thermal spalling will be resisted and restricted to progress through the refractory, and failure will be delayed. Hence, a strong refractory will fail at a higher strained conditions and will result in better thermal shock resistance.

d. Elastic modulus: During thermal cycles, refractory will be prone to change its dimension as per the ambient/outside thermal conditions. Now, if a refractory has a high value of elastic modulus, then the requirement for sudden and repeated dimensional changes caused by the thermal conditions of each cycle will be less accommodated by the refractory, resulting in cracking and failure of the refractory.

e. Porosity: For any dense refractory presence of porosity helps in increasing thermal shock resistance. Closed pores that are present in the interior of the refractory, actually act as crack arrestor/stress absorber by blunting the crack (generated by thermal shock) tip during its progress through the refractory. The sharp tip of the crack when interacts with the closed pores become blunt and directionless, and requiring much higher level of strain energy to propagate further. Thus higher presence of closed pores in dense refractory improves the thermal shock resistance.

Again, due to thermal strain (developed due to thermal shock), the cracks generate from the surface of the refractory as it faces the maximum variation in thermal conditions. Any crack that is getting generated from the surface of a refractory will have less energy to form if the refractory has higher apparent (surface) porosity, as the strain energy available from the strain of thermal shock to initiate the crack will be distributed among the surface pores (who act as a crack generator), and energy available for each pore to generate the crack will be less. Thus, a higher level of total strain energy will be required to initiate the crack, resulting in higher thermal shock resistance. Also, porosity can absorb the expansion in volume due to increase in temperature, causing a reduction in thermal strain and an increase in thermal shock resistances.

But, these porosity factors are valid only for the dense refractories. For insulating refractories, the porosity level is quiet high, which results in the poor strength of the refractories. Hence, higher porosity in insulating (nondense) refractory is actually of no benefit, and increase in porosity in a porous body will further deteriorate the strength property and also the thermal shock resistance.

The thermal strain also gets generated in a refractory sample if a huge temperature gradient exists between its two surfaces. The dimension of the hot face will be higher than that of the cold one, and if thermal expansion coefficient is high thermal conductivity is low for the material, the strain generated will be enormous. Again, this strain will keep on increasing with the increase in cyclic temperature change or prolonged use, resulting in failure of the material.

3.9.4　Permanent Linear Change on Reheating (PLCR)

This is another typical property especially required for the refractories. It can be visualized as the dimensional change (in general shrinkage) of the refractory on the second firing. Shrinkage (discussed in Section 3.2.3) is the dimensional change of a refractory that occurs due to firing (from dried condition). As PLCR is the shrinkage value of the second firing, this property may be considered as the general category of physical properties. For the refractory manufacturing, it is important to obtain the desired dimensional size of the product. But, for the user, it is important to know whether there will be any further (above acceptable limit) permanent dimension change of the refractory during its use. Any dimensional change, say shrinkage, at high temperature that may occur during use of the refractory will cause a gap between the refractories, resulting in heat leakage and many other associated problems. This may also cause a collapse of the refractory structure. Again, any expansion during application may cause strain within the refractory structure, causing instability and collapse of the refractory lining. During use of the refractory, shrinkage may occur due to poor densification/sintering, and expansion may be due to any incomplete reaction within the refractory composition, if those are not completed during firing of the refractory manufacturing process.

Hence, if any fired refractory, ready for use at the application site, is again fired close to its application or firing temperature, it should have nearly no dimensional change. Any change in dimension of the refractory that occurs due to this second firing (reheating) is termed as permanent linear change on reheating (PLCR). If PLCR value is above the acceptable limit (which is very minimum), the whole refractory lot will be rejected. If the refractory manufacturer uses lower temperature of firing or lesser dwelling/soaking time during firing (to reduce the total firing cost), poor sintering or densification and incomplete reactions (if any) will occur, resulting in high PLCR values. Hence, PLCR values are important for refractory applications, especially from the users' point of view.

SUMMARY OF THE CHAPTER

Various multiapplication areas of refractory demand different kinds of properties. For dense refractories, the following properties are required:

- High bulk density (BD), low porosity (AP), low permeability, low shrinkage, and low specific gravity.
- Good mechanical properties like cold crushing strength (CCS) and cold modulus of rupture (CMOR).
- Low thermal expansion property is desired.
- High thermal conductivity for better thermal shock resistance, or low thermal conductivity is required for better heat retention.
- High thermomechanical properties, like hot modulus of rupture (HMOR) and creep.
- High resistances against wear, abrasion, and chemical attack.
- High refractoriness (or PCE), refractoriness under load (RUL), thermal spalling (or shock) resistance, and low permanent linear change on reheating (PLCR).

Again, for better performance in insulating refractory, it should have high AP, low thermal conductivity, high mechanical and thermomechanical properties, and other properties similar to dense refractories are desired (though for any refractory with high porosity the strength, hot strength, wear, chemical resistances, etc., are poor).

QUESTIONS AND ASSIGNMENTS

1. Why we need different types of properties for refractories?
2. Briefly describe the properties required for dense refractories and insulating refractories.
3. Discuss the importance of thermal conductivity of refractories.
4. What the different physical properties, and why they are important?
5. Why are thermomechanical properties more important for refractories than conventional mechanical properties?
6. Describe the requirements for resistance to chemical attack.
7. Why wear properties are required for refractories?
8. Detail the difference among HMOR, Creep, and RUL. Discuss which property is more appropriate to importance and relevance of each of these properties.
9. What do you understand by thermal shock resistance? Describe the refractory properties on which thermal shock resistance is dependent and how?
10. What is PLCR? How is it different from shrinkage? Why is it important?
11. Why carbon monoxide disintegration test is important for fireclay refractories?
12. Why the dimensions of the sample for measuring mechanical properties are important?

BIBLIOGRAPHY

1. C. A. Schacht, *Refractories Handbook*, CRC Press, Boca Raton, US, 2004.
2. J. H. Chesters, *Refractories- Production and Properties*, Woodhead Publishing Ltd, Cambridge, 2006.
3. P. P. Budnikov, *The Technology of Ceramics and Refractories*, 4th Ed., Translated by Scripta Technica, Edward Arnold, The MIT Press, 2003.
4. *Harbison-Walker Handbook of Refractory Practice*, Harbison-Walker Refractories Company, Moon Township, PA, 2005.
5. Stephen C. Carniglia and Gordon L. Barna, *Handbook of Industrial Refractories Technology: Principles, Types, Properties, and Applications*, Noyes Publications, Westwood, NJ, 1992.
6. A. Rashid Chesti, *Refractories: Manufacture, Properties and Applications*, Prentice-Hall of India, New Delhi, 1986.
7. W. David Kingery, H. K. Bowen, and Donald R. Uhlmann, *Introduction to Ceramics*, 2nd Ed., John Wiley & Sons Inc, New York, NY, 1976.
8. Felix Singer and Sonja S. Singer, *Industrial Ceramics*, Springer, Dordrecht, Netherlands, 1963.
9. F. H. Norton, *Refractories*, 4th Ed., McGraw-Hill, New York, US, 1968.
10. W. E. Lee and W. M. Rainforth, *Ceramic Microstructures: Property Control by Processing*, Chapman & Hall, Sheffield, UK, 1994 [ISBN: 978-0412431401].
11. W. D. Kingery, H. K. Bowen and D. R. Uhlmann, *Introduction to Ceramics*, 2nd Ed., Wiley India Pvt Ltd, Delhi, India, 2012 [ISBN: 978-8126539994].

4 Standards and Testing

4.1 INTRODUCTION TO DIFFERENT STANDARDS

Refractories are industrial materials used for all the high-temperature processings where a selection of proper refractory for a particular application is important. This selection is done as per suitability of the refractory for that application that can meet all the required property criteria of the application environment with the economy. Each application area demands a specific set of properties that may be unique to that application only.

Most of the properties of a refractory are quantitative in nature (other than some visually descriptive ones like color, appearance, etc.). They are used as a metric by which the benefits of one type are measured and assessed by another type, and proper refractory selection is done for a particular application. These properties are functions of independent variables, like temperature, direction of the measurement, testing environment, etc. Some of the properties are dependent on the sample size also.

Refractories are classic examples of ceramics that contain preexisting pores, flaws, and cracks in them due to processing and manufacturing techniques. These deformities are not uniform among the batches of the same composition and even among the different samples of the same batch. So the properties of each testing of the same batch may vary, and an average of few samples is a must for accuracy and dependability in the results obtained.

Again for any application, not a single property is important rather a set of properties are required to be satisfied by the refractory, and the measurement techniques of all these properties must be acceptable to all with accuracy and repeatability. So every measurement of each testing has to be done in a precise and predetermined fixed conditions that will result in exactly similar values in each and every time of testing. So a set of fixed parameters is required for carrying out any testing or evaluation of the property, which will become the "standards" for that testing.

A "standard" is a document that provides requirements, specifications, guidelines, or characteristics that can be used consistently to ensure that materials, products, processes, and services are fit for the intended purpose. It details the conditions and parameters for any testing, represents an indispensable level of information in that specific area, established by consensus mainly of the manufacturer, user, researchers, government agencies, and consumers. It is also certified and accepted by a recognized government organization that provides rules and guidelines for common and repeated use. Any standard is a collective work of different forum of people to meet the demands of the society and technology of better living.

The use of standards is becoming a prerequisite for trade, especially in the case of global business. A very large percentage of export is influenced by the international standards business. Above all, everyone can benefit from the conformity and integrity that standards will bring out. Use of standards helps our daily lives in many ways,

DOI: 10.1201/9781003227854-4

making life easier, safer, and healthier by identifying and classifying products conforming to particular requirements, and warrants service performance in long run.

Most of the technologically strong countries are having their standards. Most common and globally accepted standards that are used for refractories are

- British Standards Institutions of UK, code "BS."
- American Standards for Testing Materials of US, code "ASTM."
- Conformite Europeene (European Conformity) of European Economic Area, code "CE."
- International Organization for Standardization (ISO) standards of various countries, code "ISO."
- Deutsches Institut für Normung (German Institute for Standardization) of Germany, code "DIN."
- Association Française de Normalisation of France, code "AFNOR."
- Japanese Industrial Standards of Japan, code "JIS."
- GB (Guobiao) Standards of China, code "GB."
- Bureau of Indian Standards of India, code "IS."

All the different standards are having their method of testing refractory materials; in many cases, the parameters are same. The refractory material must follow the norms and satisfy the condition and criteria of the standards of a particular country or global zone to do business in that area.

4.2 TESTING OF REFRACTORIES

In the previous chapter, we have seen various types of refractory properties and the different subclasses of each type to characterize a refractory material. Each of the class and subclasses is important for different refractories from the application points of view. Evaluation of these different classes and subclasses of properties is done as per certain specific standards. There are different norms mentioned in the different testing standards prevailed in the different parts of the world. Here, in the present chapter, the testing methods are described in a much-generalized way mostly for the shaped and fired refractories, covering mainly the popular and most widely used properties and the standards. The most common properties that are important and used for refractories have been discussed in Chapter 3, and the testing methods to evaluate them are discussed below.

4.3 TESTING OF PHYSICAL PROPERTIES

4.3.1 Bulk Density (BD), Apparent Porosity (AP), Water Absorption (WA), and Apparent Specific Gravity (ASG)

These properties are commonly used to describe and judge a fired shaped refractory, like, brick. Also, they are used for comparing the same quality of refractory obtained from different source/supplier to judge the quality for selection purpose in a specific application. Methods used to measure these properties are primary standard methods suitable for quality control, research and development activity, comparison and selection, compliance with specifications, etc. purposes. Common conditions and assumptions associated with the test methods are,

1. samples must not react with water (otherwise, nonaqueous liquid is to be used),
2. maximum extent of original surfaces of the fired shaped article are retained in the test samples,
3. requirements regarding the size, configuration are met,
4. sample are not very fragile or having loose grains that may fall apart,
5. surface pore are fully impregnated by boiling or vacuum method, etc.

If any of the above stated conditions and assumptions are not fulfilled, the test results may be erroneous.

The presence of porosity in refractories does not allow to measure the density and other related properties by simple Archimedes Principle, as used for nonporous samples. Hence, a modified Archimedes method is used for porous samples. The sample size required to measure these properties varies from standards to standards. Table 4.1 details the sample sizes mentioned in different popular standards and used commonly. There are two main methods used for the determination of the above properties, which are

a. Water boiling method, and
b. Vacuum or evacuation method

4.3.1.1 Boiling Method

The test specimen is cut from the main refractory, retaining as much fired surfaces as possible. Then the sample is dried in an oven at 110°C, and, after reaching a constant mass, dried weight (D) is taken. The sample is then immersed completely in distilled water as suspended condition such that none of the sample surfaces (even the bottom) touches the surface of the water container. The container, containing water and the immersed suspended samples, is then heated, and the water is allowed to boil for 2 hours. The sample should not touch the container walls also during the boiling period. If the water level goes down due to evaporation, the addition of water is to be done so that the samples must remain as immersed condition all through the boiling

---

TABLE 4.1

Sample Size for BD and AP Study as per Different Standard Specifications

Specification Standard	Sample Size
ASTM C 20 - 00 (reapproved 2015)	One quarter of a brick retaining four original surfaces
JIS 2205	One half or one quarter of a normal brick
IS 1528 part XV: 2007	65 mm × 65 mm × 40 mm

---

period. After 2 hours of boiling, the container along with the suspended samples is kept away from the heat source and allowed to cool down. After cooling, the suspended weight (S) of the sample is taken. This weight is taken while the sample remains in suspending condition in water with the help of a loop or halter (made up of copper wire) hung from a balance. The balance must be counter-balanced with the wire in place and immersed in water to the same depth as is used when the test sample is in place.

Next, the sample is taken out from water, and the extra surface water present is wiped off lightly from the surface by blotting with a wet/moistened towel/smooth linen or cotton cloth to remove all drops of water from the surface. Care must be taken such that blotting/wiping is only enough to remove the excess surface water present on the surface. Excessive blotting/pressed wiping can produce an error by withdrawing water from the surface pores of the specimen. After wiping, the sample is again weighed for its soaked weight (W).

This W corresponds to the weight of the sample when the surface pores are filled with water. Hence, it is equal to the dry weight of the sample plus weight of water present in the surface pores. Now the term (W-S) indicates the mass difference of the sample measured in air and immersed in water when the surface pores are filled with water. Removal of air from the surface pores has occurred during boiling as the volume of air present increased due to heating and came out of the surface and sample continuously during boiling, which volume was replaced by the water present nearby during the cooling period. Thus $(W-S)$ indicates the bulk volume (without the volume of the surface pores) of the sample in CGS unit, as per Archimedes Principle. Hence,

$$\text{Bulk density (BD)} = (\text{mass/bulk volume}) = D/(W-S)$$

Again the term $(W-D)$ shows the weight of water present only in the surface pore of the sample. Hence, in CGS unit, the value of $(W-D)$ is the volume of the water present in the surface pores that is the volume of the total surface pores. Hence,

Apparent porosity (AP) (%) = volume of surface pores/bulk volume = (W–D) × 100/ (W–S).

Water absorption (WA) is defined as the percentage of water absorbed by the sample per unit mass of it. So,

$$WA = (W-D) \times 100/D$$

Apparent specific gravity (ASG) is defined as $D/(D-S)$

4.3.1.2   Vacuum/Evacuation Method

In this method, the measurement parameters and measuring conditions are exactly the same as that of the boiling method. Only the difference is the process part of the experiment. Here, the samples are kept in vacuum as immersed and suspended condition in water instead of placing them for boiling. Any organic liquid (like kerosene, xylene, etc.) can be used instead of water. Organic liquids cannot be used in boiling methods due to their high evaporation rate. For samples having hydration tendency or react with water, organic liquids have to be used instead of water, and only vacuum method can be used for these measurements. Samples are immersed in liquid without touching the container surfaces (similar to boiling method), and the whole system is placed inside an empty vacuum desiccator, which is then evacuated to a vacuum level of 2.0 kPa (~25 mm Hg column or 0.02 atmospheric pressure). This vacuum level is maintained for at least 2 h, after which air is allowed to enter, and the measurements are done exactly in a similar way as mentioned in the boiling method. In this technique, the vacuum draws out the air from the surface pores, and spaces are filled with the surrounding liquid. If a liquid other than water is used, then the density of the liquid is required to be multiplied for calculation of BD and ASG.

4.3.2 TRUE SPECIFIC GRAVITY (TSG) AND TRUE DENSITY (TD)

True specific gravity is the ratio of true density, determined at a specific temperature, to the density of water at that temperature and has no unit. It is a primary property related to chemical and mineralogical composition of the sample. Specific gravity helps to classify the refractories, can detect the differences in chemical composition and mineralogical phases or phase changes, also essentially required to calculate the total porosity and closed porosity of a refractory sample.

The test is done by immersing the powdered sample in a liquid, mostly water, if the sample has no reaction with water. Few assumptions are also associated with the measurement process. The assumption is: (i) sample used is a true representative of a bulk material (though only few gms are used), (ii) nearly no porosity is present below the fineness of the powdered sample, (iii) no impurity has been introduced during the preparation of the sample, etc. If the assumptions are not strictly valid, then the results may not be accurate.

First a representative piece of the refractory sample is to be taken. A higher number of samples may lead to better average value and accuracy in results. Each sample is to be crushed and ground first, and then sieved through at least 150-micron sieve. No selective grinding and no exclusion of any portions that may difficult to crush or grind is to be done. Particles passing through the sieve are demagnetized to remove any magnetic (iron) particles contaminated during crushing and grinding process. The powders were then collected, and sampling is done to reduce the amount of sample for the test by coning and quartering method.

The powdered sample (about 50 g) is dried at 110°C and poured into a special glass bottle with stopper rod named pycnometer bottle [weight of the empty bottle W1(in g)], and then the bottle containing the sample is weighed (W2 in g). Next the bottle is filled with water (other liquid) in such a way so that the whole powder is immersed in liquid, and the bottle is filled to 1/2 or 2/3 of its capacity. Then, it is placed inside a vacuum desiccator and vacuum is started. Bubbles will start coming out from the sample through the liquid. A vacuum level of less than 25 m bar is maintained. With time, the removal of air bubbles will decrease, and then it will die down. The vacuum is maintained for 1–2 hours. The pycnometer bottle with the sample and liquid inside is slowly shaken intermittently during the vacuum process to make sure of complete wetting of the material by the liquid. Instead of a vacuum, boiling can also be used, as used for BD and AP study, when water is used as the liquid medium.

Next, the pycnometer is taken out from the vacuum desiccator, and the sample particles are allowed to settle down if disturbed during vacuum or boiling process. Then the bottle is filled with the liquid, and the glass-stopper rod is placed. Care is necessarily required for filling and placing of glass-stopper so that there will not be any overflow of sample powder out of the bottle. The capillary of the stopper rod must be filled with the liquid. Any excess liquid coming out from the tip of the stopper rod or outside of the rod and outside of the bottle is wiped off with care so that it does not draw out any liquid from the capillary of the rod. Next, the bottle with rod containing the sample and completely filled with liquid is weighed (W3 in g). Then the pycnometer is emptied completely, washed, and dried. The clean and dried bottle is then filled with the test liquid, and the stopper rod is placed in such a way that the capillary inside the rod is filled as before. The weight of the filled pycnometer is taken (W4 in g).

True specific gravity and true density are then calculated as per the following formula, and the results are reported to the nearest third decimal place,

$$\text{True specific gravity (TSG)} = (W_2 - W_1) \times D_l / [(W_4 - W_1) - (W_3 - W_2)], \text{ and}$$

$$\text{True density (TD)} = (W_2 - W_1) \times (D_l - D_a) / [(W_4 - W_1) - (W_3 - W_2)] \text{ g/cc}$$

where, D_l and D_a represent the densities of the liquid used and air, respectively, at the test temperature.

The ratio of bulk density to the true density represents the fraction of densification that has occurred in the sample. The ratio can be used to have an idea about the densification of the sample during sintering. Percentage of this fraction represents the percent relative densification of the sample.

$$\text{Percent Relative Densification} = (BD/TD) \times 100\%$$

Again, the value $(1 - BD/TD)$ represents the volume fraction that has not densified or the fraction of total porosity present in the sample. Hence,

$$\text{Total porosity} = (1 - BD/TD) \times 100\%$$

Now, this total porosity has two components: one is the surface or apparent porosity, the measurement process is described in Section 4.3.1; and the other one is closed porosity, which remains within the sample and cannot be assessed or measured directly. Hence, this closed porosity can be calculated as total porosity – apparent porosity. So,

$$\text{Closed porosity} = [\{1 - BD/TD\} \times 100 - \text{apparent porosity}]\%$$

The sample preparation techniques used in this testing that is crushing and grinding only makes the sample finer than 150 microns (or the sieve size used), and the sample is not necessarily free from the closed pores that are finer than this size. The amount of residual closed pores may vary between the refractories and even within the samples of the same refractory. And the process mentioned above does not consider these fine pores. Hence, the values generated by the weights of different readings are close approximations rather than accurate representations of true specific gravity values. So for any comparison purpose and any refractory selection purpose, the results obtained by this method must be judiciously judged due the presence of these fine closed pores and differences in accuracy that may generate for the same.

4.3.3 PERMANENT LINER CHANGE ON REHEATING (PLCR)

Permanent linear change on reheating (PLCR) represents the permanent dimensional change of a fired refractory shape that may occur on the second firing (reheating). Refractory shapes of different compositions exhibit unique permanent linear changes after reheating, which is primarily important for the refractory users. This test method provides a standard procedure for heating various classes of already fired refractories with appropriate heating schedules and measuring the dimensional changes. Dimensional changes may vary from batch to batch of the same source of material with the same composition of refractory due to processing inconsistency. And it is important to have an idea about the dimensional changes of the refractory during use.

Again the reheating that is done with a specific heating schedule for a certain firing time may not be comparable to the actual application conditions. Also, the heating environment may vary from the actual applications. So the permanent change that may occur during application may differ from the testing results. But, it helps to compare among different classes and subclasses of refractories, and to select the most suitable one for the specific application area. The measurement is essentially required for the refractory user industries and also for the developmental activities. Selection of representative sample for the test is important.

The different specification is having different dimensions for this test, as shown in Table 4.2, but all the specifications mention to retain as many original/molded surfaces as possible. If the original refractory shape is smaller than the required dimensions, then the largest possible size is cut from that shape ensuring that the structure of the refractory is not damaged. Dimensions and volume of the samples are measured for the purpose of calculation. Then, the samples are placed in the furnace in such a way that they are rested on the larger surface above a supporting brick (of the same refractory quality or, at least, similar refractoriness in quality). Nonreactive suitable refractory grains are placed in between the test samples and the supporting brick, and also among the test samples. A specific gap (~40 mm) is also maintained between the samples. Furnace atmosphere needs to be oxidizing, and the

---

TABLE 4.2

Sample Size for PLCR as per Different Standard Specifications

Specification Standard	Sample Size
ASTM C 113-14 (reapproved 2014)	228 mm × 114 mm × 64 or 76 mm
JIS R 2208	114 mm × 20 mm × 20 mm
IS 1528 Part VI	50 mm × 50 mm × 125 mm

flame must not impinge on the sample surface. The furnace hearth should have a uniform temperature all-through, which is to be measured by calibrated thermocouples, or by calibrated optical or radiation pyrometer. The final reheating temperature is dependent on the type of refractory that is being evaluated and its refractoriness. Every standard specification has some specific heating schedule, and the highest temperature of firing is about 50°C–100°C lower than the firing temperature of the refractory during manufacturing. After this firing for PLCR, the sample is allowed to cool down within the closed furnace and taken out from the furnace after cooling. Any small blister formed on the surface of the samples is removed by rubbing with abrasive blocks, and the dimensions and volume are measured again. The PLCR is measured in percentage, as

$$(L_f - L_i) \times 100/L_i - \text{for linear change} \qquad \text{and}$$

$$(V_f - V_i) \times 100/V_i - \text{for volumetric change}$$

where L_f and V_f are the linear dimension and volume of the sample after reheating, respectively, and L_i and V_i are that of the before reheating firing.

4.4 TESTING OF MECHANICAL PROPERTIES

Mechanical properties are required to understand the strength of a material and essentially measured for the refractories in any application. The major two strength measurement techniques at ambient temperature are discussed below.

4.4.1 COLD CRUSHING STRENGTH (CCS)

CCS values give an idea about of the suitability of a refractory for use in a specific application, only from the load-bearing capacity point of view. That is again only for room temperature conditions and provides no idea about the performance at elevated temperatures. The results obtained from this testing must be judiciously used for comparison purpose, as it depends on sample dimensions and shape, the nature of the testing surfaces (original fired, sawed, or ground), the orientation of the surfaces during testing, the loading pattern and rate, etc. Strength values may vary from source to source, batch to batch, and even within the same batch of a specific quality of refractory.

The test specimen size varies from standards to standards, and few are mentioned in Table 4.3. Test samples of required size are cut from larger refractory shapes retaining the maximum number of original surfaces (molded and fired). If the original shape of the refractory is smaller in dimension than the standard size required, the maximum size possible in cube or cuboid form is required to be used. In such cases, only one specimen shall be cut from a single fired shape without weakening or damaging the refractory. Samples with cracks or other visible defects must be rejected for testing.

The sample must be completely dried and then cooled, and the surfaces of the sample that will receive the load from the testing machine shall be ground or cut to plane and parallel. If required, a sand–cement mortar or plaster of paris paste is to be used to fill up any depression mark on the loading surface to make it perfectly

TABLE 4.3

Sample Size for CCS as per Different Standard Specifications

Specification Standard	Sample Size
ASTM C 133-97 (reapproved 2015)	51 mm cube or cylinder with 51 mm diameter and height
JIS R 2206	60 mm cube
IS 1528 Part IV	230 mm standard brick or 75 mm cube

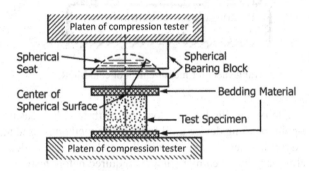

FIGURE 4.1 Testing arrangement for measurement of CCS.

flat. Asbestos fibreboards or cardboards of 5 mm in thickness and extending the sample size over the edges are to be used as bedding material at the bottom and top of the specimen. After placing the sample in the compression testing equipment with the bedding material, the load is applied (Figure 4.1). A spherical bearing block (lubricated for easy and accurate adjustment for proper placement) is to be used at the top of the test specimen in contact with the top surface and must be in the vertical axis of the test sample. The load is to be applied uniformly during the test. Loading is done to maintain a strain rate of 1.3 mm/minute for both the dense and porous (insulating) refractories with a corresponding stress rate 12 MPa/minute for dense and 3 MPa/minute for insulating refractories. At a certain load the sample will crush (break)/collapse, and the maximum applied load value is noted. The cold crushing strength is reported as calculated by the following formula:

$$\text{Cold crushing strength} = W/A$$

where W = maximum load at the moment of crushing or breaking in kgf and A = average of the gross areas of top and bottom of the sample in m^2. The size of the test sample and the loading direction applied is to be mentioned in the test report.

4.4.2 Cold Modulus of Rupture (Cold MOR)

Cold MOR test is performed on bar samples using a standard mechanical or hydraulic compression testing machine at a constant rate of stress increase. The maximum

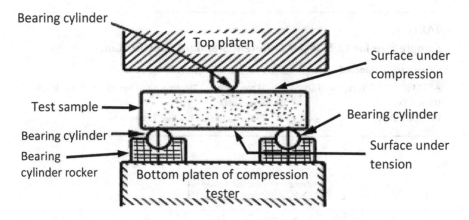

FIGURE 4.2 Testing arrangement for measurement of cold MOR.

stress that a sample of specified dimensions can withstand is measured when it is bent in a three-point bending device until failure. The method is used mainly for the shaped and fired refractories; for any chemically bonded, or tar-bonded, or unshaped products, a preliminary heat treatment may be required before testing.

The loading set up must have three bearing edges: two to support the test piece and one for the loading (Figure 4.2). The three edges have a specific radius of curvature as per the sample dimensions and have a length more than 5 mm greater than that of the breadth of the test sample. The contact lines of the edges must be parallel to one other in a direction perpendicular to the length and the plane of the breadth of the test piece. The loading device must apply the load uniformly across the center of the sample and increase it at a uniform rate. The maximum load at the time of failure is recorded.

Each test piece shall be a whole standard rectangular brick [230 (228) mm × 115 (114) mm × 75 (76) mm or 230 (228) mm × 115 (114) mm × 65 (64) mm] or any one of the sizes 200 mm × 40 mm × 40 mm or 150 (152) mm × 25 mm × 25 mm. If test pieces are cut out from fired shapes, then the cutting must be done in such a way that the loading direction for testing must match with the pressing direction of the shape.

Test samples are first dried in an oven at 110°C and then cooled, avoiding any moisture absorption. Next, the breadth and height of each test piece are measured. Samples are then placed on the lower-bearing edges of the loading equipment so that it rests symmetrically on two supporting (bearing) cylinders. When the test piece is a normal standard brick, the face bearing any brand mark, i.e. the upper face, shall be in compression. If the test pieces have been cut out of the brick, the face of the test piece that corresponds to the original face of the brick (if it has been preserved) shall be in compression. Next, the load is applied vertically to the test piece until failure occurs. The rate of increase load must conform to

a. a stress rate of 9 MPa/minute for a dense shaped refractory (corresponding to a strain rate of 1.3 mm per minute) to be maintained,

b. a stress rate of 3 MPa/minute for a shaped insulating (porous) refractory (corresponding to a strain rate of 1.3 mm per minute) to be maintained.

The maximum load at which failure of the test piece occurs is recorded and the test temperature is noted. Repeatability and reproducibility of the test are to be checked.

4.5 TESTING OF THERMAL PROPERTIES

Measurement of thermal properties for refractories is important as their application is always at high temperatures. Other than conventional thermal properties, like thermal expansion and thermal conductivity, refractoriness or PCE of a refractory material is also considered as thermal properties.

4.5.1 REVERSIBLE THERMAL EXPANSION

Reversible thermal expansion (RTE) is measured by dial gauge method using a horizontal-type or vertical-type dilatometer. A dilatometer is an equipment containing a refractory tube, which is closed at one end and open on the other; a refractory rod that holds and supports the test sample during testing; and a programmable furnace that can cover both the tube and the rod with calibrated thermocouple. The test sample is placed in a slot within the tube, rest between the closed end of the tube and the rod. Generally, the tube and the rod are made up of same ceramic/refractory material. The open end of the rod is connected through a spring support to a dial gauge system (generally graduated in divisions of 0.01 mm) or digitally connected for the measurement of exact dimension. After placing the sample, the dial gauge or digital display reading is adjusted to "zero" value to measure any change in dimension, both expansion and shrinkage. Next, the whole system, that is, the closed end of the tube, the sample, and the rod, is placed inside a furnace. Figure 4.3 shows the dilatometer and the sample holding system.

The test specimen (as a refractory cylinder) has the dimensions of 50 mm length and 10 mm diameter, and is obtained from the refractory shape by core drilling or by cutting and grinding. The end faces of the sample must be very flat and parallel to each other. For a test temperature up to 1150°C RTE is measured in a dilatometer (generally vertical) with fused silica tube and fused silica rod. On an increase in temperature, any increase (or change) in the dimension of the sample will force the rod, and the change in dimension will be indicated by the micrometer dial gauge or digitally displayed. The temperature is measured by a thermocouple placed inside the furnace very close to the sample. When this temperature of thermal expansion measurement is higher than 1150°C, the vitreous silica system may crack due to the devitrifying effect within itself; high alumina system (mainly recrystallized alumina set ups) is used, and horizontal dilatometer is used.

Calibration of the apparatus is done before the experiment to check and measure any temperature variation between the outside of the refractory tube and the specimen and the expansion in length of the tube. A 50 mm standard piece (of fused silica or alumina as per the system) is used for calibration, and any deviation from the zero reading on the dial gauge is noted against the temperature. The expansion values obtained in all subsequent tests with the apparatus shall be corrected by adjusting the difference between the dial gage reading at a given temperature during the calibration test and the true expansion of 50 mm standard piece at that temperature.

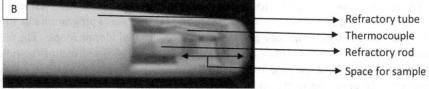

FIGURE 4.3 (a) Dilatometer equipment and (b) Sample holding space in the tube of dilatometer.

For the test sample, the length is measured accurately by slide calipers and then is placed at the closed end of the tube in between the two refractory disks (to avoid any displacement during the test). Next, refractory rod is placed on the free end of the disk on top of this specimen. A proper thermocouple is then inserted into the tube in such a way that the hot junction is positioned almost at the center of the specimen and is connected to a temperature recording device. Next, the dial gauge is adjusted for zero reading. The whole system with the test sample is then placed in the furnace, and heating is done at a predetermined rate (as per the specification), generally 5°C per min. The temperature and the corresponding dial gage along with the reading of the dial gauge is noted at a regular interval, may be 5 minutes.

The dial gauge reading is corrected according to calibration data, and the corrected values represent the linear dimensional change of the sample with the change in temperature. It is expressed as a percentage of the original length of the specimen. A graph relating to percent expansion value against temperature is also plotted for graphical representation.

4.5.2 THERMAL CONDUCTIVITY

The thermal conductivity of refractories is important as it determines the heat transfer character. This is important as the user industries select the refractories to attain a specified (precalculated) conditions of heat loss and corresponding cold face temperature, without exceeding the temperature limitation of the refractory used. So it is essential to know the heat conduction behavior of the refractory, heat loss, and amount heat to be supplied to continue and complete the high-temperature process. If a composite lining (multiple layers of refractory) is used, the thermal conductivity values of all the individual lining refractory are important to calculate the heat balance of the process. Different methods are used for measuring the thermal conductivity of refractories, as described below.

4.5.2.1 Calorimetric Method

This test method covers the determination of thermal conductivity of refractories under standardized conditions of testing using a calorimeter. It requires a large thermal gradient and steady state heat transmission conditions. The results are based on a mean temperature and are suitable for making refractory specification, selection of a refractory for a particular application, designing of refractory lining, construction of furnaces, etc. The use of these data requires consideration of the actual application environment and conditions. A special apparatus is used for this method, having:

1. an electrical heating chamber up to a maximum temperature of 1540°C in a neutral or oxidizing atmosphere and can maintain uniform heat distribution,
2. a copper calorimeter assembly that measures the amount of heat flowing through the test specimen by the water circulation process,
3. a water-circulating system with regulator valve to provide constant water supply to the calorimeter assembly at constant pressure and temperature conditions. The water is at ambient temperature with a minimum pressure of 30 kPa,
4. calibrated thermocouples with hot junctions embedded in the test specimen and the cold junctions immersed in a mixture of ice and water,
5. a multiple differential thermocouples for measuring the temperature rise of the water flowing through the calorimeter.

Three numbers of standard bricks {230 (228) mm × 115 mm × 75 mm} and six numbers of soap bricks [230 (228) mm × 64 mm × 57 mm] with uniformity in structure and bulk density (measured after drying at 110°C), free of broken corners or edges, are used for the test. One straight brick is used as the test specimen, and one each of the other two brick is used as guard brick on either side of the specimen. The six soap bricks are placed around the edges of the test specimen and guard bricks to prevent any side heat flow. The test specimen and guard bricks cover an area of approximately 456 mm × 342 mm.

The 230 (228) mm × 115 mm faces of the three straight bricks and the 230 (228) mm × 64 mm faces of the soap bricks are ground to make them flat and parallel. The sides that are to be placed in contact shall be ground flat, and at right angles to the 230 (228) mm × 115 mm face of the straight brick and the 230 (228) mm × 64 mm face of the soap brick.

The following data are recorded for each 2-hour test period (steady state of heat flow):

1. linear dimensions of the test specimen,

2. distance between thermocouple junctions located in the test specimen,
3. three sets of temperature readings, as measured by the thermocouples in the test specimen,
4. mean temperature between each pair of thermocouples in the test specimen, as calculated from the temperatures recorded in °C,
5. average rise in temperature of the water flowing through the calorimeter,
6. average rate of water flow through the calorimeter, and
7. the rate of heat flows through the test specimen per unit area.

Thermal conductivity is calculated as

$$k = qL/[A(T_1 - T_2)]$$

where k=thermal conductivity, W/m·K, q=rate of heat (W) flowing into the calorimeter [temperature rise in K of the water flowing through the calorimeter times the weight of flowing water kg/s], L=thickness (distance between hot junctions at which T_1 and T_2 are measured, in m), T_1=higher of two temperatures measured in the test specimen (in K), T_2=lower of two temperatures measured in the test specimen (in K), and A=area of center calorimeter (m²).

4.5.2.2 Parallel Hot-Wire Method

It is a dynamic thermal conductivity measurement technique for refractories based on the increase in temperature of a certain location placed at a specified distance from a linear heat source and embedded between two test pieces. The test pieces are heated in a furnace to the desired temperature; then local heating is done by passing electricity through a wire with known power embedded in the test piece along the length of the test piece. A thermocouple is fitted at a specified distance from the hot wire and measures the increase in temperature as a function of time. From all these data thermal conductivity is calculated.

The apparatus used for this technique consists of:

1. an electrically heated furnace that can heat the test assembly up to 1250°C
2. a hot wire, platinum or platinum/rhodium, about 200 mm in length and maximum 0.5 mm in diameter. One end of the wire is attached to the lead for the supply of the heating current, and the other end is attached to a lead for measurement of voltage drop
3. a stabilized A.C. power supply to provide current to the hot wire (preferably constant power supply)
4. platinum/platinum–rhodium thermocouple (Type R or S) formed from a measurement thermocouple and a reference thermocouple connected in opposition
5. a data acquisition system of temperature–time registration device with a sensitivity of at least 2 pV/cm and a temperature measurement to 0.05 K.

Each test assembly consists of two identical test pieces with standard brick dimensions or at least 200 mm × 100 mm × 50 mm in size. The surfaces of the test pieces

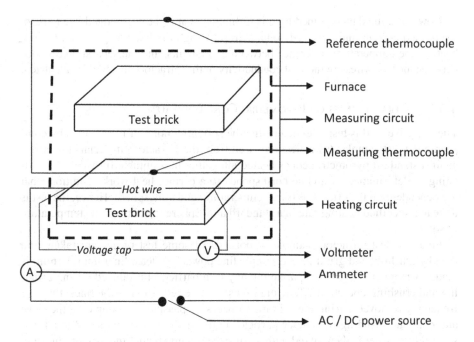

FIGURE 4.4 Schematic diagram of experimental setup for parallel hot-wire method.

that are in contact with each other must be very flat. A groove to accommodate the hot wire and the thermocouple shall be machined in both the contact faces.

The hot wire and differential thermocouple are placed between the two test pieces, with the hot wire along the center line of the brick faces in contact with each other. Next, the test assembly is placed inside the furnace, resting each assembly on support of the same material for uniform heating. Then the test assembly is connected to the measuring apparatus with the hot-wire circuit open. The schematic diagram of the testing setup is shown in Figure 4.4.

Next the furnace is heated to the test temperature at a maximum of 10 K/minute. After reaching the test temperature, a soaking of 10 minutes is given for uniformity in temperature. Then the heating circuit is closed, and a fixed power input (chosen as per recorder sensitivity) is provided in the hot-wire circuit. The exact moment of initiation of the power supply is noted, and the voltage drop across the hot wire and the current in it immediately after switching on the heating circuit and again at intervals during the test period are measured and noted.

After an appropriate heating time, the heating circuit is disconnected. All measurements and recordings of the differential thermocouple are stopped. Allow time for the hot wire and test assembly to reach temperature equilibrium, and verify the uniformity and constancy of the temperature. Repeat the procedures from closing the heating circuit and power input and temperature measurements, so, obtaining a further measurement of the rate of rising of the temperature of the hot wire under the same conditions.

Now, from the data obtained by this technique, that is, heat generated due to passing a fixed current under a fixed voltage drop through the fixed length wire, time of heating, the temperature difference between two circuits, and the separation distance between the two wires the thermal conductivity of the refractory material is calculated.

4.5.3 REFRACTORINESS OR PYROMETRIC CONE EQUIVALENT

The objective of this test is to determine the softening point of refractories by comparing the test samples (as cones) prepared from the refractory materials under test against standard pyrometric cones heated altogether in a suitable furnace. The softening or deformation of a cone corresponds to a certain heat-work conditions that are dependent on the effects of time, temperature, and atmosphere. Hence, the test is done at a specified heating rate in a standard atmosphere to compare the temperature of softening.

First, the test refractory sample or portions of some test pieces are taken, and then by crushing and grinding the sample fine powder (at least finer than 0.2 mm) is made. A magnet is used to separate out any iron particles introduced during grinding and crushing operations. To avoid excessive size reduction of the fines, they are frequently removed during the grinding process by passing the sample on the sieve and continue grinding of coarser particles until all the sample passes through the sieve. The powder is then mixed with alkali-free organic binder (dextrin or glue) and water. Test cones are prepared by placing the mixture in a metal mold, preferably of brass, in the shape of truncated trigonal pyramid with dimensions of 8 mm on the sides of the base and about 25 mm high, as shown in Figure 4.5. Sufficient handling

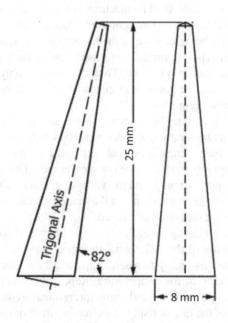

FIGURE 4.5   Schematic diagram of pyrometric cone.

strength in the green shapes is developed by drying or by a preliminary firing at about 1000°C.

The cones made from the refractory test sample are mounted along with the standard pyrometric cones (Seger/Orton/equivalent) on a refractory round plaque with the help of a bonding material. Plaque and the bonding material must not react and affect the test and standard cones. Mounting of the test and standard cones are done with about 3 mm deep embedment in the plaque and one of the faces inclined toward the center of plaque at an angle of 82° with the horizontal. Test cones and standard cones are placed at the outer edge of the plaque, and arranged in such a way to place test cones in between the standard cones. Standard cone are selected with some anticipated range, as far as practicable.

Next, the plaque containing the cone samples are placed inside a special furnace (PCE furnace) and is heated, following some standard specific schedule, initially very fast, reaching about 1600°C in about 1 hour. Necessary care is required to have a uniform distribution of heat, to avoid any direct flame contact with the samples and to ensure oxidizing atmosphere during firing. The cones are regularly viewed to check their condition during the whole heating process through the top opening or a peep hole. At higher temperatures, the cones will start to soften, and the softening temperature of a cone is marked by the bending over of the cone and the tip of the cone touching the plaque surface (Figure 4.6). Any bloating, squatting, or unequal fusion of small constituent particles are noted. Once the test cones soften, the tip of the cone will bend toward the inclined direction (center of the plaque) and touch the surface of the plaque on increasing temperature. Once the test cones touch the plaque

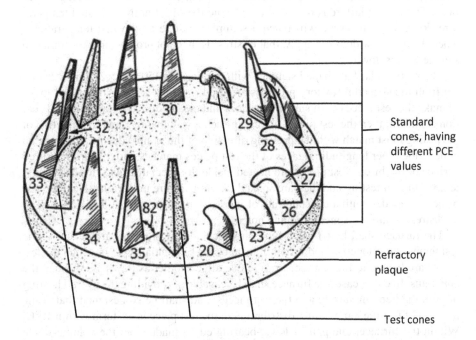

FIGURE 4.6 Schematic diagram of conditions of the pyrometric cones during the test.

surface on bending, the heating and the experiment is stopped. The softening point of the test sample is reported with reference to the standard pyrometric cones with similar softening behavior. The standard cone number that nearly corresponds to the softening behavior of the test cone is the equivalent cone for the test sample and its number is the PCE value. If the test cone softens in between two standard cones and approximately midway between, the softening point is reported by mentioning both the cone numbers like cone No. 35–36. If the test cone starts bending at an early cone but it does not bend down completely, or its tip does not touch the base even until a later cone, the same needs to be reported.

4.6 TESTING OF THERMOMECHANICAL PROPERTIES

Evaluation of the thermomechanical properties of the refractories is of greater importance from the application points of view. Any refractory has to withstand high load at elevated temperatures, and their strength evaluation is of prime importance in that condition.

4.6.1 HOT MOR

Evaluation of the modulus of rupture of refractories at elevated temperatures has become a widely accepted means to evaluate and judge materials performance at service temperatures. A progressive application of force or stress is done on a specimen rested on two supports (as a simple beam) close to two ends and a center-point loading inside a furnace with uniform elevated temperature under oxidizing conditions. The load at failure is recorded to evaluate the MOR at that elevated temperature. Refractory materials will reach a semiplastic state at elevated temperatures, where Hooke's law does not apply, that is, stress is then not proportional to strain and sample breaks down.

The rectangular bar-shaped samples with dimensions 150 mm × 25 mm × 25 mm cut from the shaped refractory products are used for this testing. Attention is given to make the test pieces with smooth surfaces and clean edges. If the pressing direction is known, then the test pieces are cut in such a way that the loading direction of the testing must match with the pressing direction of the sample during shaping, and none of the other longitudinal faces of the test pieces matches with the original fired surface of the brick. Test pieces are heated inside the special furnace to the specific temperature of testing and then are soaked for temperature uniformity. Next the test pieces are loaded with a constant rate of increase in stress until failure occurs. The load/stress at failure is measured and noted.

The furnace shall be (i) batch type, in which some test pieces are heated to the test temperature together and tested in turn, or (ii) sequential type, in which the test pieces are heated to the test temperature one after another as they pass through the apparatus. In either case, the furnace shall be capable of providing the overall heating of both the bending device and the test pieces, and shall be so designed that at the moment of test the temperature distribution in the test piece is uniform within 10°C. Within the furnace, one pair of lower-bearing edges made from the volume-stable refractory material is installed with a gap of 125 mm at centers. A thrust column, containing the top bearing edge that is made from volume-stable refractory, is also extended to outside the furnace where means are provided for applying a load. The lower-bearing edges and the bearing end of the support column shall have rounded bearing surfaces having about the 6 mm radius. The thrust column is maintained in vertical alignment, and all bearing surfaces are parallel in both horizontal directions. The atmosphere inside the furnace is air (oxidizing environment). The temperature is measured by a calibrated thermocouple in the proximity of the mid-point of the tensile face of the test piece.

A heating rate of 2°C–10°C per minute is maintained with a soaking time of 30 minutes at the test temperature. After the soaking period, loading is done on the sample with a loading rate of,

a. 0.15 N.mm^{-2}.s^{-1}, for a dense-shaped refractory product, and
b. 0.05 N.mm^{-2}.s^{-1}, for an insulating refractory product.

Once a test sample is broken, the final load is noted. The next sample is then moved along the lower-bearing edge, and placed under the trust column and break them by the preceding procedure. From the sample dimensions, the gap between the supporting bearing edges and load value, hot MOR is calculated.

4.6.2 CREEP

Creep can be defined as the isothermal deformation of a stressed product as a function of time. Here, a test sample of given dimensions is heated under specified conditions to a given temperature, and a constant compressive load is applied to it. The deformation of the test sample, as the percent change in dimension (length) of constant temperature and load, is recorded as a function of time.

The test piece is a cylinder of 50 mm diameter and 50 mm height, with a hole from 12 to 13 mm in diameter, extending throughout the height of the test piece, bored coaxially with the cylindrical outer surface. The top and bottom faces of the test piece must be plane and parallel by sawing (and grinding if necessary), and is perpendicular to the axis of the cylinder. The surface of the cylinder shall be free from any visible defects.

The loading device is capable of applying a load centered on the common axis of the loading column (moving one), the test piece and the supporting column (fixed column), and directed vertically along this axis at all stages of the test. The sample is rested between the loading column and supporting column with the same common axis. A constant compressive load is applied in a downward direction from above on the piece resting directly or indirectly on a fixed base. Any deformation of the test piece is measured generally by a dial gauge with a sensitivity of measuring of 0.005 mm. Two disks, 5–10 mm thick and at least 50 mm in diameter of an appropriate refractory material compatible with the test material, are placed between the test piece and the loading and supporting columns. The columns and disks are capable of withstanding the applied load up to the final test temperature without significant deformation. Also, there should be no reaction between the disks and the loading system.

A vertical axis furnace, with a capacity to raise the temperature uniformly to the desired test temperature by 5 K per minute in air atmosphere, is used for the test. During any period of constant temperature, the fluctuations of temperature must be controlled within 5 K. A central thermocouple passing through the fixed column and the sample through the axis is placed at the mid-point of the test piece, for measuring the temperature of the test piece at its geometric center. A control thermocouple, placed in a sheath and situated very close to the test piece, is used to regulate the rate of rising of temperature. The thermocouples shall be made from platinum and platinum–rhodium wire, and shall be compatible with the final test temperature.

Precisely measured test sample is placed between the supporting and loading columns with the spacing disks, and the measuring device is adjusted to the correct setting. Next, a constant compressive load is applied to the loading column at one or other of the following stages in the test:

    a. At the moment when the furnace is switched on, i.e., from ambient temperature;
    b. After the test piece has been soaked at the test temperature for a given time (minimum 1 hour, maximum 4 hours).

This loading condition is to be stated in the test report. The specific load applied to the test piece may vary from experiment to experiment, and the total load applied to the sample must include the mass of the moving column and the associated disk for any calculation. A specific load of 0.2 N/mm^2 (0.2 MPa or 2 kg/cm^2) for dense refractories and 0.05 N/mm^2 for insulating products are commonly used. The total load used is rounded to the nearest 1 N value. For the test where the load is applied when the furnace is switched on, recording of the changes in the height of the test piece and its temperature is done at a gap of 5 minutes during heating and for the first hour after attaining constant temperature as indicated by the control thermocouple. After that record the changes at every 30-minute intervals till the completion of the test.

When the load is applied to the test piece soaked at the desired constant temperature, recording of the change in height of the test piece and its temperature is done by starting with load application at a 5 minutes gap for the first hour and after that at every 30 minutes gap till the completion of the test. The standard duration for testing ranges from 25 to 100 hours. In a test where the load is applied when the furnace is switched on, the results are plotted as percentage change in the height of the test piece as a function of temperature first till the attainment of the desired temperature, and then it is plotted against the time of test at that fixed temperature.

4.6.3 REFRACTORINESS UNDER LOAD

Refractoriness under load (RUL) indicates the maximum applicable temperature for any refractory and determines the softening temperature of the refractories under a specific load indicated either by complete sloughing down or breaking of the test specimen. A load of 2 kgf/cm^2 (0.2 MPa) is applied on the cylindrical sample with diameter and height both of 50 mm. The cylindrical test sample is obtained by boring or cutting and grinding out of the central portion of the brick to be tested. The original surface of the brick should form one of the end faces of the finished test specimen. Specimens with cracks or other visible defects must not be used.

The furnace used for this testing is a specially made cylindrical electrically heated one using coke particles to conduct the high-voltage power supply and generate heat from their resistance. Coke particles form the outer layer of the furnace, and packing of the coke particles controls the heating rate of the furnace. In the inner portion, the furnace consists of a heating tube (corundum, magnesite, or mullite make) of 100–120 mm in diameter and about 500 mm length with a wall thickness of 10–15 mm. The zone of maximum temperature shall have a minimum length of 100–120 mm at the central portion of the tube. A loading arrangement is done to provide a constant load of 0.2 MPa, including the weight of the loading rod, vertically to the test piece. Provision is also made for recording the changes in the height of the test specimen during heating by a dial gauge and to permit it to be compressed by at least 20 mm. The set up for the test is shown in Figure 4.7.

Specific load of 0.2 MPa is applied to the test sample after it is placed in-between two rods (support rod and loading rod) generally made up of carbon or mullite. Carbon plates of 5 mm thickness and above 50 mm diameter are used in between the rods and the test specimen. The dial gage reading, for measuring the dimensional change of the sample, is adjusted to zero position, and then heating is started. The specific heating rate is followed as per different standard specification. As per Japanese JIS R 2209, it is 6°C per minute till 1000°C and 4°C per minute above 1000°C; as per Indian IS 1528 Part 2, heating rate is 15°C per minute up to 1000°C and above 1000°C at a constant rate of 8°C per minute. The difference between the actual temperature rise and the scheduled rise of temperature should not vary more

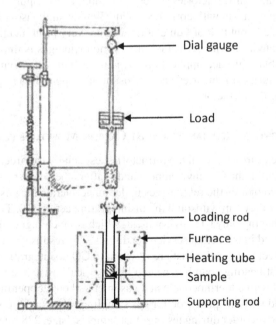

FIGURE 4.7 Schematic diagram of the set up for RUL test.

than 20°C at any point of time. The dimensional change of the sample is plotted, and a temperature versus deformation curve is obtained. The temperature is measured with an optical pyrometer, sighted or adjusted upon the bottom of a refractory tube closed at its bottom and suspended in the furnace at the beginning of the test at about the middle of the test specimen. As per JIS 2209, temperature corresponding the maximum increase in length of the test sample is marked as the apparent initial softening temperature (T_1), and also the temperatures corresponding to 2% (T_2) and 20% (T_3) shrinkage in height of the samples are noted, and the temperature ranges $(T_2 - T_1)$ and $(T_3 - T_1)$ are calculated. As per Indian standard, (IS 1528) temperature $(T_a,$ termed as the temperature of appearance) corresponding to 0.6% deformation and temperature $(T_e,$ termed as the temperature of the end) corresponding to 40% shrinkage is noted and reported. In the case of premature breaking of the test specimen, actual softening does not take place, and the temperature of breaking (T_b) is reported.

As per ASTM C 832 (reaffirmed 2015), the similar testing is termed as "thermal expansion under load." In this testing, a sample of 38 mm × 38 mm cross section with 114 mm height is used under a load of 172 kPa stress, and a heating rate of 55°C ± 5°C per hour is maintained. Continuous linear change data is recorded in a computerized system, or manual system, after every 55°C till 1095°C and then after every 28°C interval. Heating is stopped when the linear expansion is ceased, and the temperature is noted. This temperature corresponds to the maximum level of expansion when the creep (deformation) rate equals to the expansion rate.

4.7 TESTING FOR CORROSION RESISTANCE

Chemical attack on refractories is very common, as the applications of the refractories involve contacts with corrosive solids, liquids, and gases. The corrosion from solid is very less, but that of liquid and gases are very prominent. Among the different gaseous environments, effect of carbon monoxide gas is important for refractories used in blast furnace applications. Again for liquid environments, corrosion of slag in metallurgical industries and corrosion of molten glass in glass industries are important for the refractories.

4.7.1 TESTING OF RESISTANCE AGAINST CARBON MONOXIDE (CO)

The presence of iron oxide in the refractories, especially for fireclays, causes a disintegration effect in the CO environment and shatters the whole refractory lining. This test method compares the refractories under an accelerated exposure to CO to determine whether they can withstand the disintegrating action CO. The test is done in an atmosphere having only CO environment, which is much higher in the concentration of CO prevailed in the actual service conditions. The results obtained by this method is used to select refractories that are resistant to CO disintegration.

A gas-tight heating chamber, heated by electrical resistance wire, is used for the test provided with a thermocouple and gas inlet and outlet openings. The chamber is also equipped with the provision for gas sampling at the outlet port and temperature controller and recorder during testing. Test samples, [size, 228 mm × 64 mm × 64 mm or 228 mm × 76 mm × 76 mm, ten in number (as per ASTM C 288) or 50 mm long

30 mm dia cylindrical sample, two in number, (IS 1528 Pt 13)] with as many original surfaces as possible, are made by cutting from fired refractory shapes. For unfired shapes, a preliminary firing at around 1100°C is necessary for the test pieces before the test.

First, the heating chamber containing the test specimens is heated to an operating temperature of 495°C–505°C (ASTM) or 450°C (IS) in a nitrogen atmosphere. After attaining the temperature, the atmosphere of the chamber is changed by passing CO gas through the inlet and drained out from the outlet opening. The heating chamber should contain at least 95% CO. Carbon monoxide gas used for the test is taken from a gas cylinder, or is generated from formic acid and sulphuric acid, or by passing the carbon dioxide from a cylinder over charcoal heated around 1000°C in a tube furnace. The CO gas is purified by passing it through a purifying train to remove carbon dioxide, oxygen, and water vapor. Magnesium perchlorate, silica gel, or phosphorus pentoxide are also used for the removal of water vapors. A manometer and flowmeter are included in the gas train before the entrance of the gas into the furnace. A constant flow rate of CO [2 in^3/hour·in^3 (ASTM) or 2 L/hour (IS)] is maintained all through the testing. After specifically selected period, the samples are inspected. Before each inspection, a fast flow of nitrogen through the furnace is done to flush out the CO, and then a slow flow of nitrogen is maintained during cooling. Inspection is done for general discolouration, carbon deposition, cracking, and disintegration. The test is continued for a maximum of 100 hours or until the test pieces disintegrate, whichever is earlier. Only the period during which the hot test pieces are exposed to the stream of carbon monoxide is considered.

Reporting of the test results is done by mentioning the state of the test samples after testing.

a. Unaffected: when no particles spall with no cracking, and nearly no visible change is observed.
b. Affected: surface popouts, when destructive action is confined to spalls or surface popouts of up to 13 mm in diameter.
c. Affected: cracked, when destructive action produces spalls or popouts greater than 13 mm in diameter, or cracking, or both.
d. Destructive condition: when the specimen breaks into two or more pieces, or when hand pressure can cause breaking.

4.7.2 TESTING OF RESISTANCE AGAINST LIQUID CORROSION

Among different liquids that affect and corrode the refractories, slag and glass are most common. So the measurement of corrosion against these liquids is necessary and important for selecting the suitable refractory for a particular application. But the application conditions of a refractory vary widely and are complex in nature. Hence, the standardization of this corrosion test is difficult to replicate the conditions that are prevailed in the actual application environment. Different test methods are practiced as per suitability and easiness to evaluate the corrosion of the refractory. Unfortunately, most of the testing methods for liquid corrosion are not available in the standard specifications, and mostly the testing are done as per conventional or

customized practices. The most commonly used and easy method for testing corrosion against a liquid is static cup method; there are dynamic testing methods also,
namely, finger test and rotary slag tester, which are described in the following.

4.7.2.1 Static Cup Method

In this method, a cylindrical hole is made at the central portion of a refractory cube
or cuboid, retaining sufficient thickness of the refractory wall in all the side including
bottom. Next, the solidified powdered sample of the corroding material (slag or glass)
is placed in the hole. Then the refractory sample with the corrosive powder is heated
in a furnace to the desired temperature, generally up to the application temperature
of the refractory and hold there for some prefixed time period. At that test temperature, the corrosive powder will melt and react with the refractory surfaces in contact.
Corrosion will occur in the refractory surface portions in contact with the corrosive
liquid. After cooling, the samples are cut through the vertical axis, and the dimension
of the corroded (eaten away) portion is measured. Also, any component/constituent
of the corrosive liquid may enter (penetrate) within the refractory but may not vis-
ibly corrode it. In such a case, the penetration of any particular component within
the refractory is measured as penetration depth. A schematic diagram of the axially
cut corroded sample is shown in Figure 4.8. Also, the detailed microstructural study
of the corroded portion of the samples may be carried out to understand the reactions that might have occurred, components of slag and refractory that have reacted,
details on the formation of liquid phase, any appearance of new phase/compound, or
disappearance of exiting phase, etc.

But this testing method is not replicating the actual conditions where the refractory surface and the corrosive liquid are in dynamic contact, and the refractory surface is always in contact with the fresh liquid. Static condition of testing and contact
of the refractory with the same fixed amount of slag (causing saturation of slag)
reduce the severity of the corrosive action. These do not represent the actual situation
of refractory application, and the cup method of corrosion testing is less severe than

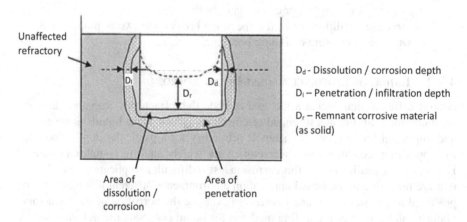

Unaffected refractory

D_d - Dissolution / corrosion depth

D_l – Penetration / infiltration depth

D_r – Remnant corrosive material (as solid)

Area of dissolution / corrosion

Area of penetration

FIGURE 4.8 Schematic diagram of the cross section of the cup after static corrosion test.

the actual refractory applications. However, through the static method, idea about corrosion of the refractory, trend of reaction, the reacting phases that are involved in the corrosion, and the product phases that are formed, etc. are obtained. This test method may be used for comparison purpose among different types of refractories against a specific corrosive liquid composition.

4.7.2.2 Finger Test (Dynamic Method)

This corrosion testing method involves of suspending small pieces of refractory samples, bar, or cylinder (like a finger shape) in a corrosive liquid (slag or glass) bath. This method may also be termed as immersion method and is used when the quantity of the corrosive is far more than refractory. In most of the cases, this test is done on slag samples. The refractory sample is rotated within the molten slag bath for a specific period, thus imparting a relative motion (dynamic effect) between the refractory surface and the reacting slag. Also, the refractory is facing the fresh slag, and the slag composition remains nearly unaltered as the slag amount is much higher compared to refractory sample. (But for static method, as the corroded components of the refractory get dissolve in the limited amount of slag, the slag composition changes and corrosiveness got reduced.) After the test, the suspending sample is taken out of the slag, allowed to cool down, cut through the cross section, and checked for dimensional loss as the measurement for corrosion. Also the penetration depth for any component of slag, if any, can be measured. Detailed microstructural studies can also be conducted, as mentioned in a static method.

4.7.2.3 Rotary Slag Test (Dynamic Method)

Rotary slag test evaluates the slag corrosion behavior of refractories in a rotating furnace, generally in the presence of a reference material. This testing method allows comparing the corrosion behavior of different refractory samples, also a standard material, under the same testing conditions against a specific slag composition. The obtained results are useful for the development of new products or in the selection of refractories to be used in contact with similar slag compositions.

The rotary slag tester consists of a rotating furnace placed in little tilted position, made up of a cylindrical metal shell mounted on rollers, and is motor driven. A gas burner capable of heating to very high temperature, say 1750°C, attached with gas and oxygen flowmeters is used. The slag is constantly charged and renewed to maintain the original slag composition and to keep similarity as that of the application conditions. The addition of fresh slag automatically removes the already reacted molten slag present in the slag pool of the tester, and a slag flow occurs in the downwards direction. The flow of the slag can also cause mechanical erosion of the refractory similar to the actual application areas. The tilt angle and rotational speed of the furnace are specified and fixed, as they affect the wear rate.

Test specimens of length 228 (230) mm and cross section and shape (as shown in Figure 4.9) are used for the testing with slag contact face as the original molded surface. One or more reference samples are included in each test run along with the samples. A number of test samples, six or more, are arranged to constitute the test lining. Suitable granular refractory or castable is used behind the test lining to hold the test samples with the cylindrical shell. At both the ends, the inert material lining

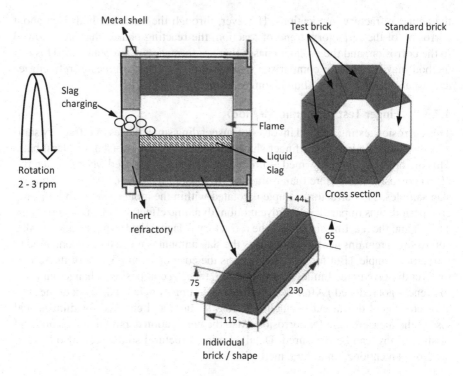

FIGURE 4.9 Schematic diagram of the rotary slag corrosion tester, including the cross section and refractory test sample shape and size.

is given to protect the shell. The whole assembly is then held in place by retaining rings bolted to each end of the shell. The shell, with the test specimens, then placed in its cradle and linkage is made to the driving motor. The gas-oxygen torch mounting is adjusted for firing axially through the furnace. Ground and shaped (by extrusion or pressing) slag pellets are fed into the slag tester and counted to determine the number of pellets charged during the test to calculate and control the slag flow rate.

The furnace is typically tilted ~3° axially down toward the burner end. The furnace, preheated by the gas-oxygen torch, is fired to a temperature to melt the slag pellets. After reaching the desired temperature, 30 minutes soaking is done. The molten slag washes over the lining and drips from the lower end of the furnace in front of the burner. Furnace rotates at a constant speed 2–3 rpm, and an oxidizing atmosphere is maintained. This rotation imparts the dynamic motion between the slag and the refractory. During the test, slag layer temperature is measured by an optical pyrometer. Regular feeding of slag pellets is done to make fresh contact with slag, and the whole process runs for about 5 hours (depending on the quality of the refractory). After the stipulated period, the firing and the rotating motor is stopped. Next, the furnace is tilted to a vertical position to remove all the remaining slag. After cooling, the furnace is disassembled, and the refractories are checked through the length perpendicular to and at the center of the slag contact face. Measurement of the dimensions for corrosion, penetration, is taken after cutting the samples, and microstructural studies, if required, are done as mentioned earlier.

4.8 TESTING OF ABRASION RESISTANCE

This test method measures the resistance of a refractory material against abrasion. Abrasion of refractory is defined as the surface wear caused by the mechanical action of moving solids. The test method determines the volume of material abraded from a flat surface of a test piece placed at right angles to a nozzle through which abrading grains are blasted by compressed air at a specific pressure. Mostly all the different standards are having the same concept of measurement, and only the individual testing parameters are different.

The test system requires an abrasion tester consisting of a Venturi blast assembly of an air blast gun and a nozzle. The nozzle directs the abrading grains onto the test piece under compressed air. There is also a feed mechanism capable of supplying a fixed amount of abrasive grains to the blast assembly at a specific time. A tightly sealed test chamber is used with a door to permit ready access for mounting and removing the test pieces. A manometer is used to measure the pressure inside the chamber during the test with a vacuum gauge to check the pressure at the entry point for the abrasive on the blast assembly. Silicon carbide abrasive grains of size between 0.85 and 0.6 mm amounting 800 g and 0.6–0.3 mm amounting 200 g (total 1000 g) is used for the testing.

Samples of dimension 100 mm × 100 mm × 25 mm are cut from refractory bricks or shapes in such a way so that one of the square faces of each test piece is an original surface. Next, the test pieces are dried to constant weight, and its bulk density is measured. Then each piece is placed in the test chamber with the square face (original molded surface) facing the nozzle at a distance of 203 mm from the nozzle mouth. Compressed air supply with regulated pressure at 450 kPa is used to maintain a chamber pressure of 310 kPa. After the desired level of pressure is obtained in the test chamber and compressed air system, the feed mechanism is connected to the blast assembly, and the total abrasive grains are blasted on the sample for 450 s. After the test, the weight of the sample was again measured, and the abrasion loss (A) is calculated as

$$A = \text{Weight loss/Bulk density (unit cc)}$$

4.9 TESTING OF THERMAL SHOCK RESISTANCE

Thermal shock or spalling is the cracking or fracture of a refractory caused by differential expansion due to thermal shock, a steep temperature gradient, a crystalline conversion due to temperature fluctuation, or a change in composition near the hot face. For qualitative measurement, expressed in number of cycles that a refractory material can withstand, different techniques may be used to determine the thermal shock (spalling) resistance. There is also some quantitative measurement technique of thermal shock resistance, measured as the loss of strength or loss of Young's modulus of any sample after some specific number of thermal cycles. The popular methods for determining the thermal shock resistance are described as below.

4.9.1 WATER QUENCHING METHOD

This is a qualitative measurement of thermal shock resistance. A refractory test specimen of standard brick dimension 230 mm × 115 mm × 75 mm is taken. An electric heating furnace with controlling facility to maintain the temperature within 10°C is used. A water tank that can maintain constant water flow is used for quenching/cooling the heated specimen.

The test specimen is dried at 110°C to constant weight, and the weight is taken. Next, the test specimen is inserted into the furnace chamber, maintaining a temperature of 1300°C, up to a depth of 50 mm lengthwise. Any gap is to be filled in by light fireclay inserts and asbestos. After insertion of the test specimen in the furnace, the temperature is again raised to 1300°C and the specimen is kept at 1300°C for 10 minutes. After heating, the specimen is taken out from the furnace, and the heated end is dipped into the water tank to a depth of 50 mm (the heated part only). The water tank is connected to running water at ambient temperature with a flow rate of 7 L/minute. The test specimen is kept in the water for 5 minutes, and then taken out and kept in air for 5 minutes. This heating for 10 minutes, and then cooling in water for 5 minutes and in air for 5 minutes completes one thermal cycle. This cycle of heating and cooling of the specimen is repeated, and any weight loss of the sample is checked after each cooling. When there is a weight loss of 20% of the specimen, lost by flaking off, the test is stopped, and the number of the cycle just before the 20% loss is reported as the number of cycles that the specimen can withstand.

4.9.2 SMALL PRISM METHOD

This is also a qualitative measurement of thermal shock resistance of a refractory sample. Test pieces are cut or ground to the shape of prisms 75 mm high with a square base of 50 mm (or rings of 50 mm height from sleeves, nozzles). A muffle or semi-muffle-type electrically heated furnace is used for heating with a thermocouple for temperature measurement.

Test samples are placed in the furnace at cold condition. Heating is done to reach the desired temperature, say 1000°C, in about 3-hours' time. First, a soaking of 30 minutes is given at that temperature for uniformity of temperature in the sample, and then the test pieces are taken out from the furnace. Hot samples are then placed on their square face over a brick floor having nearly no air flow. The test samples are cooled in air for 10 minutes and then again placed in the furnace, maintained at that desired temperature. One heating and one cooling step complete a thermal cycle. Next, each heating and cooling operation will be for 10 minutes only. After each cooling period, the sample is examined for any crack formation on the surface (visual observation). Once the crack appears on the sample surface, the test is stopped, and the number of the cycle during which the cracks first appear is reported.

4.9.3 DETERIORATION IN PROPERTY AFTER THERMAL SHOCK

This test method determines the strength loss or reduction in continuity, or both, of prism-shaped test samples after a certain number of thermal cycling. The strength loss is measured by measuring the cold modulus of rupture (MOR) of the samples before and after thermal shocks. Also, the reduction in structural continuity is estimated by the difference in sonic velocity before and after thermal cycles. Also, it indicates the ability of the refractory to withstand the stress generated by sudden changes in temperature from the heating and cooling operations.

Visually crack- and flaw-free test samples, ten in number from two different bricks or shapes, with dimensions 150 mm × 25 mm × 25 mm and having, at least, one original surface in each sample are used for this testing. An electrically heated furnace that can heat and maintain 1200°C with temperature recovery time of less than 5 minutes, a standard mechanical or hydraulic compression testing machine, and a sonic velocity measuring machine are the apparatuses required for this testing.

First, the sonic velocity along the length of each test specimen is measured, and, next, the cold modulus of rupture using three-point loading bending setup is determined. Then, the furnace is preheated to the test temperature of 1200°C, and the test specimens are placed into the furnace spanning the setter brick and kept them for 10–15 minute. Then, the samples are taken out and cooled for 10–15 minutes while spanning the setter brick in ambient air. This one heating and one cooling period are considered as one full cycle. Total five such complete cycles are conducted and each heating cycle to be considered once the furnace starts maintaining the desired temperature after recovery. After thermal cycles, the sonic velocity of each specimen is measured along the length, and cold modulus of rupture is tested. Percent sonic velocity loss of each specimen is calculated as

$$(V_i - V_f) \times 100/V_i$$

where V_i is the initial sonic velocity (m/s) of each specimen before any thermal cycle, and V_f = sonic velocity (m/s) of each specimen after specific number of thermal cycles. Also, the percent loss in modulus of rupture strength of the specimen is calculated as

$$(M_i - M_f) \times 100/M_i$$

where M_i is the average cold modulus of rupture strength of the original specimens (MPa) and M_f is the average cold modulus of rupture strength (MPa) of the specimen after specific number of thermal cycles.

SUMMARY OF THE CHAPTER

Quantitative estimation of the properties of refractories is done by various testing methods.

Testing methods for any single property may vary depending on the specific testing standard followed for the testing.

The results of the same property testing of the same material may differ as sample size, mode of testing, and the rate of certain operation involved in testing may differ in different standard specifications.

Refractories, being ceramic materials, have preexisting pores and cracks, result in nonuniformity of properties. Hence, average data of few samples is to be represented by each type of testing to get a predictable behavior of the material during use.

Different countries or zones of the globe have different standard testing methods and specifications.

Bulk density, apparent porosity, apparent specific gravity, and water absorption values can be obtained from a single experiment, which can be conducted by two different methods, namely boiling method (nonhydraulic samples) and vacuum method.

Specific gravity or true density is measured using powdered samples by pycnometric method.

Mechanical strength measurements, both at ambient and elevated temperatures, are dependent on sample size and loading rate, specification followed to measure them, etc. The results may differ from the same quality of the refractory due to them, and so all these parameters are to be mentioned in the report.

For thermal properties, thermal expansion is measured by using a dilatometer. A vitreous silica-based system is used up to a temperature of 1150°C, and alumina-based dilatometers are used for higher temperatures. Thermal conductivity is measured by calorimetric or hot-wire methods, and well precaution is required to be taken to prevent any heat loss and error in the results.

Corrosion testing can be done by various methods, but the simple techniques are far from reality and application environment. Rotary slag tester approaches close approximation to the actual application but still have issues like the rotation of refractory, noncontinuous contact with slag, etc.

Thermal shock resistance can be measured by both qualitative and quantitative methods.

QUESTIONS AND ASSIGNMENTS

1. Why we need standard specification for measuring the properties?
2. Describe in detail how bulk density and apparent porosities are measured?
3. Why do we need to consider the liquid density for calculating bulk density but not for apparent porosity when a liquid other than water is used?
4. How true specific gravity is measured?
5. How closed porosity can be measured?
6. Describe the difference between the different thermomechanical property measurements.
7. Describe the difference between different test methods for liquid corrosion resistances.
8. How refractoriness of an unknown sample can be measured?
9. Describe the difference between different test methods for thermal shock resistances.
10. Describe the measurement technique of creep.
11. Detail the testing method for thermal expansion.

12. How is carbon monoxide disintegration test done?
13. What are the drawbacks of static cup method for liquid corrosion test?
14. What is the different dynamic slag corrosion test? Detail any process of the methods.

BIBLIOGRAPHY

1. A. Rashid Chesti, *Refractories: Manufacture, Properties and Applications*, Prentice-Hall of India, New Delhi, 1986.
2. W. David Kingery, H. K. Bowen, Donald R. Uhlmann, *Introduction to Ceramics*, 2nd Ed., John Wiley & Sons Inc, New York, NY, 1976.
3. Felix Singer, Sonja S. Singer, *Industrial Ceramics*, Springer, Dordrecht, Netherlands, 1963.
4. J. H. Chesters, *Refractories- Production and Properties*, Woodhead Publishing Ltd, Cambridge, 2006.
5. P. P. Budnikov, *The Technology of Ceramics and Refractories*, 4th Ed., Translated by Scripta Technica, Edward Arnold, The MIT Press, Massachusetts, US, 2003.
6. F. H. Norton, *Refractories*, 4th Ed., McGraw-Hill, New York, US, 1968.
7. C. A. Schacht, *Refractories Handbook*, CRC Press, Boca Raton, US, 2004.
8. *Harbison-Walker Handbook of Refractory Practice*, Harbison-Walker Refractories Company, Moon Township, PA, 2005.
9. Stephen C. Carniglia, Gordon L. Barna, *Handbook of Industrial Refractories Technology: Principles, Types, Properties, and Applications*, Noyes Publications, Westwood, NJ, 1992.
10. *Standard Test Methods for Apparent Porosity, Liquid Absorption, Apparent Specific Gravity, and Bulk Density of Refractory Shapes by Vacuum Pressure*, ASTM C830 - 00, 2011.
11. *Standard Test Methods for Apparent Porosity, Water Absorption, Apparent Specific Gravity, and Bulk Density of Burned Refractory Brick and Shapes by Boiling Water*, ASTM C20 - 00, 2015.
12. *Standard Test Methods for Cold Crushing Strength and Modulus of Rupture of Refractories*, ASTM C133 - 97, 2015.
13. *Standard Test Method for Modulus of Rupture of Refractory Materials at Elevated Temperatures*, ASTM C583-15, 2021.
14. *Standard Test Method for Measuring Thermal Expansion and Creep of Refractories Under Load*, ASTM C832 - 00, 2015.
15. *Standard Test Method for Thermal Conductivity of Refractories by Hot Wire (Platinum Resistance Thermometer Technique)*, ASTM C1113/C1113M - 09, 2013.
16. *Standard Test Method for Thermal Conductivity of Refractories*, ASTM C201 - 93, 2013.
17. *Standard Test Method for Pyrometric Cone Equivalent (PCE) of Fireclay and High Alumina Refractory Materials*, ASTM C24 - 09, 2013.
18. *Standard Test Method for Abrasion Resistance of Refractory Materials at Room Temperature*, C704 / C704M -15, 2015.
19. *Standard Test Method for Disintegration of Refractories in an Atmosphere of Carbon Monoxide*, ASTM C288-87, 2014.
20. *Standard Test Method for True Specific Gravity of Refractory Materials by Water Immersion*, ASTM C135 - 96, 2015.
21. *Standard Test Method for Rotary Slag Testing of Refractory Materials*, ASTM C874-11a, 2020.
22. *Methods of Sampling and Physical Tests for Refractory Materials, Determination of Cold Crushing Strength of Dense Shaped Refractories Products, Indian Standard specification IS 1528- Part IV*, 2012.

23. *Methods of Sampling and Physical Tests for Refractory Materials, Method for Determination of Modulus of Rupture at Ambient Temperature of Dense and Insulating Shaped Refractory Products, Indian Standard Specification IS 1528- Part 5*, 2007.
24. *Methods of Sampling and Physical Tests for Refractory Materials, Determination of pyrometric Cone Equivalent (PCE) or Softening Point, Indian Standard Specification IS 1528- Part 1*, 2010.
25. *Methods of Sampling and Physical Tests for Refractory Materials, Determination of Refractoriness Under Load, Indian Standard Specification IS 1528- Part 2*, 2011.
26. *Methods of Sampling and Physical Tests for Refractory Materials, Determination of Spalling Resistance, Indian Standard Specification IS 1528- Part 3*, 2010.
27. *Methods of Sampling and Physical Tests for Refractory Materials, Determination of Permanent Linear Change after Reheating for Shaped Insulating and Dense Refractories, Indian Standard Specification IS 1528- Part 6*, 2010.
28. *Methods of Sampling and Physical Tests for Refractory Materials, Determination of Apparent Porosity, Indian Standard Specification IS 1528- Part 8*, 1974.
29. *Methods of Sampling and Physical Tests for Refractory Materials, Determination of True Density, Indian Standard Specification IS 1528- Part 9*, 2007.
30. *Methods of Sampling and Physical Tests for Refractory Materials, Method for Determination of Bulk Density and True Porosity of Shaped Insulating Refractory Products, Indian Standard Specification IS 1528- Part 12*, 2007.
31. *Methods of Sampling and Physical Tests for Refractory Materials, Determination of Resistance to Carbon Monoxide, Indian Standard Specification IS 1528- Part 13*, 2007.
32. *Methods of Sampling and Physical Tests for Refractory Materials, Method for Determination of Bulk Density, Apparent Porosity and True Porosity of Dense Shaped Refractory Products, Indian Standard Specification IS 1528- Part 15*, 2007.

5 Silica Refractories

5.1 INTRODUCTION

Silica is the name given to a specific compound composed of silicon and oxygen, the two most abundant elements in the earth's crust. Hence silica, the only compound of these two elements, is a very common and one of the most commonly available oxides in the earth's crust. Silica is found in nature commonly in the crystalline form and rarely in an amorphous state. It is composed of one atom of silicon and two atoms of oxygen, resulting in the chemical formula SiO_2. Quartz is the only form of silica that is thermodynamically stable in the atmospheric conditions, and so all the uncombined free form of silica present in the crust is available as quartz and is found in quartzite, sand, flint, and ganister. Quartzite is the most commonly available form of quartz and used as the major raw material for manufacturing silica refractories. The raw material must have a minimum amount of impurities like Al_2O_3, Fe_2O_3 and TiO_2 that greatly reduce the liquidus temperature and restricts high-temperature applications.

Silica refractories are those materials that contain about >93% SiO_2, with a minor amount of other oxides as additives and impurities, namely lime (CaO), alumina (Al_2O_3) and iron oxide (Fe_2O_3). Silica is acidic in nature and readily reacts with alkali and alkaline earth oxides, and so silica refractories are also acidic in nature. In general, two major class of silica refractories are important, namely, super duty silica refractories with very high purity level (impurity content $Al_2O_3 + Fe_2O_3$ ~0.5%) and high duty containing impurities in the range of 0.5–2 wt%. As per the historical information silica, refractories were first produced in the United Kingdom in 1822 from ganister (carboniferous sandstone) or so-called dinas sand. For this reason, in some of the countries, silica refractories are also called dinas.

5.2 RAW MATERIALS AND SOURCES

Natural sources of silica are widely and abundantly available all over the world and are used for making silica refractories. Among them, quartz is the most abundant and important. It is found in almost every type of rock: igneous, metamorphic and sedimentary. Quartz is particularly prevalent in sedimentary rocks, as it is highly resistant to break down by the physical and chemical processes of weathering. Quartz is present nearly in all mining operations, in the host rock, in the ore, as well as in the soil and surface materials called the overburden.

Silica occurs in three different crystalline polymorphic forms, namely, quartz, tridymite, and cristobalite and also as an undercooled melt called quartz glass. Each of the polymorphic forms has high- and low-temperature variations, which transform reversibly. The crystal structure of the polymorphic forms of silica differs significantly, and the transformation from one form to other is associated with a distinct

DOI: 10.1201/9781003227854-5

TABLE 5.1

Details of the Crystallographic Change of Silica

Temperature (°C)	Phase Change	Specific Gravity	Volume Changes (%)	Crystal System Change
117–163	$\alpha \leftrightarrow \beta_1 \leftrightarrow \beta_2$ Tridymite	2.26 ↔ 2.27	+0.5	Rhombic ↔ Hexagonal ↔ Hexagonal
220–280	$\alpha \leftrightarrow \beta$- cristobalite	2.33 ↔ 2.21	2.0 ↔ 2.8	Tetragonal ↔ Cubic
573	$\alpha \leftrightarrow \beta$-quartz	2.65 ↔ 2.49	0.86 ↔1.3	Trigonal ↔ Hexagonal
870	β-quartz $\leftrightarrow \beta_2$ tridymite	2.49 ↔ 2.27	14.4	Hexagonal ↔ Hexagonal
1250	β-quartz $\leftrightarrow \beta$- cristobalite	2.49 ↔ 2.21	~ 17.4	Hexagonal ↔ Cubic
1470	β_2-tridymite $\leftrightarrow \beta$- cristobalite	2.27 ↔ 2.21	~ 3	Hexagonal ↔ Cubic
1713	β-cristobalite $\leftrightarrow$ Liquid			

change in crystal structure and specific gravity. These changes are of great importance during heating and cooling of silica refractories, as associated change in volume may result in cracking and breaking of the refractory shapes.

The polymorphic forms of silica are temperature dependent, and one form changes to other on reaching a specific temperature. These changes are sluggish in nature and takes a long time to convert from one form to another due to bond breaking and reconstruction of another crystallographic structure. These changes are termed as reconstructive transformations (conversion). Again the low- and high-temperature variations of each polymorphic forms are also temperatures dependent, but as these changes are only associated with little orientational modification of the structure, the transformation is very rapid and spontaneous in nature, once the desired temperature is reached. The changes are termed as displacive transformations (inversion). As per thermodynamic stability, the principal crystalline forms of silica are quartz (trigonal and hexagonal, stable up to 870°C); tridymite (hexagonal, stable from 870°C to 1470°C); and cristobalite (cubic, stable from 1470°C to 1723°C, the melting point). Vitreous silica is a metastable phase in solid form and has relatively random network structure as that of a liquid, "frozen in" by undercooling. Table 5.1 shows the details of the transformations that occur in silica.

5.3 BRIEF OF MANUFACTURING TECHNIQUES

The selection of proper raw material, additives, and the firing process are important for the silica refractories so that the conversion of quartz to the desired form of silica, that is suitable for the intended application, can be obtained. As quartz is the naturally occurring form of silica, and polymorphic change from quartz to tridymite or cristoballite is associated with a very high volume changes, the amount of free quartz in the fired products must be minimum. Otherwise, free quartz will transform to the thermodynamically stable high-temperature form, as per the application temperature causing volumetric changes, cracking, and shattering of the refractory structure. Also, the polymorphic form of the fired silica refractory must match with the thermodynamically stable form at the temperature level of targeted applications.

The raw material for silica refractory is naturally occurring quartzite that must meet the requirements to achieve optimum brick properties. When high-temperature application is the main criteria for silica refractory, the quartzite used must have a high chemical purity with total impurity <0.5 wt%. For other applications, quartzites of ~95% purity are acceptable. Impurities like Al_2O_3, Fe_2O_3, TiO_2 and alkalis, form low-melting liquid phases and restrict the high-temperature applications.

For making the silica refractory, quartzite is first washed to remove associated clayey minerals and dust, and then crushed, ground and screened to the various grain fractions. The individual fractions are combined in predetermined proportions, according to the required application properties. In most cases, muller or pan or counter-flow mixers are used for mixing silica refractories, and different fraction sizes, namely, coarse, medium, and fines, are mixed. Bonding of silica refractories is achieved by the addition of milk of lime (hydrated lime) about up to 3 wt%. This lime acts as a mineralizer, and accelerates the formation of desired tridymite or cristoballite phase. Some more compounds namely iron oxide, borax, magnesium oxide, barium oxide and fluorides, carbonates, phosphates of sodium, lithium, and potassium also act as mineralizers. Alkalis affect the liquidus temperature very strongly. Hence, their use as mineralizers requires very strict quantity control and compositional analysis. These mineralizers are added to the silica-refractory mix in the mixer machine. Also, green binder, like cellulose, sulfate lye, molasses, etc., is added to provide green (dried) strength of the refractories after shaping (pressing). Uniform distribution and mixing of these additions in the green refractory mixture ensure the uniform and improved properties of the fired refractories. The addition of zircon and silicon carbide is also done during mixing for special cases where increased resistance to abrasion and high thermal conductivity are required in the fired refractories, respectively.

Moisture is added to the mixture to make it a semidry mix. Water acts as a plasticizer for the mixture, and helps to provide and retain the shape. The amount of moisture to be added depends on the shaping process. Table 5.2 gives an indicative idea about the amount of moisture required for different shaping processes. Higher moisture content may increase the chance of slumping of the shape after shaping and

TABLE 5.2
Variation of Moisture Requirement with Shaping Process

Shaping Process	Moisture Level (%)
Hand molding	8–10
Pneumatic ramming	6–7
Screw friction pressing	5–6
Toggle pressing	5–6
Hydraulic pressing	4–5

increase the probability of cracking during drying due to higher removal water vapor. Again, lower moisture content demands higher pressing pressure. Low-pressure shaping processes, like hand molding, produce products with low and nonniform density and strength, but can attain critical shape and much bigger sizes that are difficult to attain in machines, like in hydraulic presses. But low pressure involved in hand-molding process results in low and non-uniform density and strength values in the refractories. On the other hand, pressing process is advantageous for uniformity and improved properties with much higher productivity. But very high pressing pressure results in high compactness in the green shape, is disadvantageous for silica refractories, as it provides an insufficient gap to accommodate the volumetric expansions that are associated with the polymorphic phase transformations of silica, and may cause cracking.

After pressing, the shapes are dried to remove the physically absorbed (added) moisture, and drying time depends on the moisture content and shaping process. Drying is done up to a maximum temperature of 150°C–200°C using the waste heat of flue gas coming out from the firing kiln. For hydraulic pressed products drying time is about 24–30 hours, whereas for hand mold shapes, containing a high amount of moisture, drying time is up to 100 hours; slow drying for long duration is required to avoid any crack formation due to sudden and excessive removal of water vapor.

The firing of silica refractories requires special attention due to its polymorphic changes during heating. For this reason, the firing is done mostly in batch types of kilns with a long firing schedule, though they are less heat efficient. Continuous firing kilns require exceptionally long preheating and cooling zone to accommodate the slow heating and cooling rates to take care of the polymorphic changes, especially the quartz transformations. The $\alpha \leftrightarrow \beta$ quartz transformation, both during heating and cooling (between 550°C and 600°C), requires very slow rate to structurally accommodate the sudden and instantaneous volume change, otherwise that may generate crack. A tentative heating and cooling schedule for silica refractory, practiced in industry manufacturing process, in a batch-fired kiln is given in Table 5.3.

TABLE 5.3
Details of the Firing Schedule of Silica Refractory in a Batch-Type Kiln

Heating		Cooling	
Temperature (°C)	Period (hour)	Temperature (°C)	Period (hour)
R.T. – 500	24	Close Condition	
500–650	15	1430–800	22
650–900	17	800–600	16
900–1100	10	600–300	30
1100–1300	25	300–150	30
1300–1350	12	Open Condition	24
1350–1410	30	300–150	
1410–1430/1480	20–24		
Holding at 1430/1480	48–72		

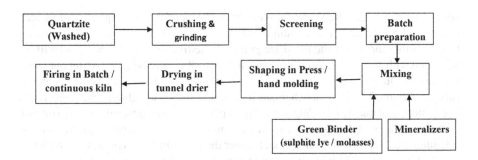

FIGURE 5.1 Manufacturing process for silica refractory.

Firing temperature of silica refractories varies between 1420°C and 1500°C, depending upon the desired phase in the final fired refractories. For the applications where tridymite phase is stable, firing is done between 1420°C and 1430°C to maximize the tridymite phase formation. For high-temperature applications of silica refractories where cristobalite phase is stable, firing is done between 1480°C and 1500°C to maximize the cristobalite phase formation. It is necessary to maintain a carefully planned time temperature schedule for getting a strong, well-bonded fired refractory with the desired polymorphic form. In general, during firing, the net linear growth of silica refractories is about 4%, and sufficient internal gaps are required within the shapes to allow such an expansion. This growth is less than the actual calculated values, because the pores present within the refractory shapes accommodate certain extent of the expansion within them. A schematic diagram for manufacturing of silica refractories is shown in Figure 5.1.

5.4 ACTION OF MINERALIZER

The chemical compounds added to accelerate and stabilize the desired mineral phase (of silica refractory) are called mineralizers. They help to mineralize the silica to a specific, desired mineral phase. Mainly lime (CaO) in the range of 2–3 wt% (added as hydrated lime) is used for making cristobalite-based silica refractories and lime ~2 wt%, and iron oxide ~0.5 wt% is used to stabilize the tridymite phase. A well-established concept is available for the formation and stabilization of cristobalite phase due to the addition of lime, as proposed by Dale, and named as the Dales theory. But for tridymite formation, such concept or theory is yet to be established.

As per the Dales theory, hydrated lime, the source of lime, losses its physical moisture within 200°C, and around 400°C, it starts decomposing to form anhydrous lime. Increasing temperature to about 800°C initiates the reaction between the fine and nascent lime with fine particles of silica in the matrix phase of the refractories, and increases the bonding within matrix phase and an increase in strength. The reaction produces $2CaO\ SiO_2$ and little free $CaO\ SiO_2$ phases. Further increase in temperature increases the reaction of fine silica with the Ca–silicate phases, and liquid phase starts forming in the CaO–SiO_2 system at about 1250°C–1300°C in the presence of other impurities like Fe_2O_3, Al_2O_3, TiO_2, etc., in the matrix phase of the silica refractory. This liquid reacts with the surfaces of the large quartz grains and dissolves the surface. More amount of silica comes to the liquid phase, as temperature is further increased and the silica grains are getting dissolved. The liquid will be saturated with silica after a certain time, and then silica will be crystalizing out from the super saturated liquid phase and will crystallize as cristobalite, as per temperature conditions. The basic mechanism involved in the cristobalite phase conversion is dissolution of quartz in a $CaO-SiO_2$ liquid phase and precipitation (crystallization) of cristobalite crystals (particles) from the liquid. This conversion of quartz to cristobalite in the presence of lime is much faster than the direct transformation with the effect of heat only, which is very slow and sluggish in nature.

5.5   CLASSIFICATION AND PROPERTIES

Silica refractories are classified mainly into two types depending on the purity or impurity content. The super duty silica refractories are termed to those who are about 97% pure and contain impurities (other than CaO, which is additive) up to the maximum of 0.5%. These impurities like Al_2O_3, Fe_2O_3 and TiO_2 are responsible for low-temperature liquid phase formation in silica, and, in general, they are termed as flux factor in silica refractories. The other type is a general-purpose silica refractory, which is about 94% pure and has impurity level up to about 2.5%. There is also one more type, called semi silica refractory, containing about 65%–80% silica, but they are of less importance and generally come under fireclay refractories.

The main features of silica refractories are:

- Low price
- Relatively low density and specific gravity
- High strength even at temperatures close to melting point (PCE)
- Nearly no shrinkage even after prolong use
- Very high corrosion resistance against acidic liquids/environments
- High thermal expansion at lower temperatures and low thermal expansion at high temperatures.

The fired silica brick contains the crystalline forms of silica, mainly cristobalite or tridymite, and little amount of unconverted and unwanted residual quartz. Additive and impurity phases present in the composition remain in the matrix as very small quantities of calcium ferrite, hematite, magnetite, alumino-silicate, etc. depending on the mineralizers used and impurities present. These minute amount of crystalline phases are responsible for the coloration and spot formation on the surface of the fired products. Silica refractories with the identical chemical composition can have differing mineralogical compositions, and this can cause quite a different behavior during use. Therefore, it is not always sufficient to evaluate silica bricks solely by their chemical composition, but essentially by the degree of transformation of quartz (or the residual quartz content) and the final phases and their content in the fired refractories.

Free quartz present in a fired silica refractory is termed as residual quartz (RQ), and, for any application, a higher RQ value is detrimental. Hence, determination of

RQ value for any silica refractory is essential. This can easily be determined by the quantitative phase analysis study through x-ray diffraction techniques. Also, accurate determination of specific gravity of the fired product gives well idea about the residual quartz content. A high RQ value results in instantaneous volumetric changes during temperature change at 573°C due to α to β quartz inversion and also will convert to equilibrium-stable tridymite or cristobalite phases depending on application temperature with huge volumetric expansions. These expansions may result in cracking and collapse of the refractory structure. An RQ value of less than 10% is essential. During use of the silica refractory, conversion of the residual quartz occurs at high temperatures, and RQ value reduces with prolonging heating or with increasing number of heating cycles.

Silica refractories are most important for their acid-resistance properties. Silica remains unaffected by any acidic conditions and retains its structural integrity, except for hydrofluoric acid and phosphoric acid above 400°C. Compounds of alkalis and alkaline earth elements attack silica particularly at high temperatures and form low-temperature-fusible silicate compounds. So silica refractories are never used in any basic environment.

Heating up and cooling down of silica refractories from room temperature to 600°C needs special attention for their high thermal expansion values and chances of cracking and disintegration due to inversion of quartz. Sudden changes in length are caused by the transformation behavior of the space lattice of silica associated with the structural changes with increasing temperature. It is also to be noted, in particular, that the thermal expansion value of cristobalite is considerably greater than that of the tridymite. Thermal expansion behavior of the different crystallographic forms of silica is shown against temperature in Figure 5.2. Again, quartz shows such a transformation at 573°C, tridymite at 117°C and 163°C, and cristobalite between 220°C and 280°C. Due to these phase changes and high thermal expansion values,

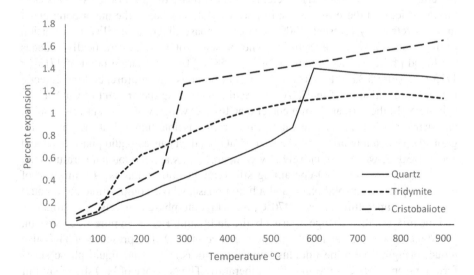

FIGURE 5.2   Thermal expansion values of *different* crystallographic forms of silica.

the thermal shock resistance of silica refractories up to 600°C is low. However, when the temperature remains above 600°C, the thermal shock resistance of silica refractories is good.

The most important characteristic of silica refractories is their refractoriness under load value, which approaches very close to that of the fusion point. No other refractory material shows such a high RUL value, close to its PCE. These two values are very close, especially for the silica refractories having a little amount of impurity. This is due to the low amount of liquid phase formation. Again whatever liquid phase is forming, they are very high silica-containing liquids and have a very high viscosity. Hence, sliding of grains in the presence of high viscous liquid (deformation) under load is negligible. Thus, the refractory has nearly no deformation till the temperature reaches close to its fusion point and has very high RUL values.

5.6 EFFECT OF IMPURITIES WITH BINARY AND TERNARY PHASE DIAGRAMS

Impurities present in the raw materials react with the silica during firing of the brick and also during use. Again chemicals and ions from the environments/corrosive agents may react with the refractory and affect the properties. These impurity ions react with the main refractory system and form compounds with a completely different character or affect the high-temperature properties by forming low-melting compounds or reducing the liquidus temperature. The presence of a second ion (impurity) in a pure system starts forming a liquid phase at a lower temperature from the eutectic/peritectic/solidus temperature, sometimes at a temperature much lower than the fusion temperature of the pure system.

For silica refractories, the main secondary oxides phases present are CaO (as mineralizer), Al_2O_3, Fe_2O_3, TiO_2, etc. Hence, the phase diagram of the systems containing silica and these oxides are important. But, as silica is the major component in silica refractory (present >90% level), the phase diagram of silica rich portion is important for silica refractories. The presence of alumina drastically reduces the liquid phase formation temperature to 1587°C from a melting point of 1713°C/1723°C (Figure 5.3a). There is a sharp fall in liquidus temperature, and the eutectic is at a composition of ~5 mol% Al_2O_3. Again, increasing the amount of CaO shows (Figure 5.3b) the formation of a eutectic at 1650°C with a CaO content of ~7 mol%, and then a further increase in CaO forms an immiscible liquid. But the phase diagram shows that any minute presence of CaO in silica causes liquid phase from and above 1430°C, which is relatively low compared to its application temperatures. At that temperature, the CaO-containing silica composition remains as a mixture of tridymite phase as a solid phase and a liquid phase, whose composition and amount vary with temperature. Above 1470°C, the tridymite phase converts to cristobalite, and the mixture then change as cristobalite and liquid. Hence, in pure silica system, liquid phase will start at 1430°C in the presence of CaO. The presence of TiO_2 also reduces (Figure 5.3c) the liquidus temperature drastically, and liquid phase starts forming from 1550°C—the eutectic temperature. The presence of FeO shows similar behavior as observed for CaO (Figure 5.3d); liquid phase starts forming from the

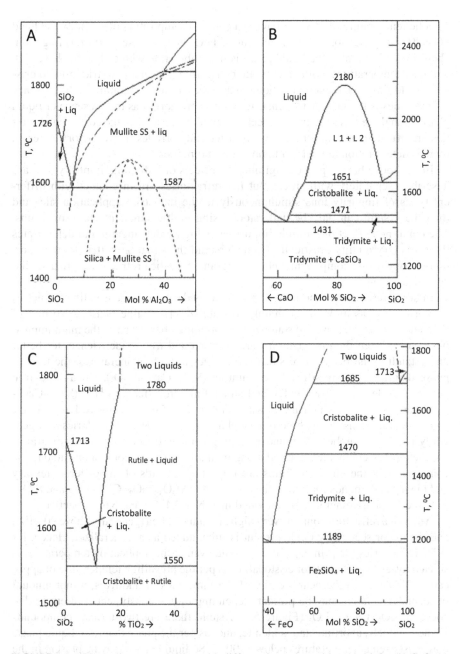

FIGURE 5.3   Phase diagrams of silica with different impurity oxides, (a) Al_2O_3, (b) CaO, (c) TiO_2, and (d) FeO.

eutectic temperature of 1685°C, having a eutectic composition of ~4 mol% FeO and 96 mol% SiO_2. But for a higher amount of FeO, liquid phase starts forming from about 1189°C. Initially, the liquid phase is in combination with tridymite phase and, at higher temperatures, with cristobalite phase after conversion of tridymite to cristobalite. Mainly, the reduction in liquid phase formation temperature in the presence of impurities is the most detrimental for silica refractories, as the presence of liquid phase strongly reduces the thermomechanical properties. Also, the corrosion properties are affected, as liquid reacts much faster and corrodes the refractory. Overall, there is deterioration in the high-temperature properties.

In the above binary phase diagrams, the effects of individual impurity ions are described separately with silica. But the refractories are in contact with different types of impurity ions simultaneously in the industrial application site, and then the system can only be understood using multicomponent phase diagrams. The combined effect of multiple impurities is highly detrimental and deteriorates the refractory quality drastically. To understand the situation better, some important ternary (three components) phase diagrams for silica refractories are also discussed as below.

In silica refractories, the most common secondary component is lime, which is primarily used as mineralizer mainly to stabilize cristobalite phase. Also mix of lime and iron oxide is used to stabilize the tridymite phase. Again, the main impurities coming from the raw materials are alumina and iron oxide. Hence, for better understanding the behavior of silica refractories at high temperatures, the ternary phase diagrams among silica–lime–alumina (SiO_2–CaO–Al_2O_3), silica–lime–iron (ferrous) oxide (SiO_2–CaO–FeO), and silica–lime–iron (ferric) oxide (SiO_2–CaO–Fe_2O_3) are important, which are shown in Figures 5.4–5.6, respectively. Each ternary phase diagram consists of three binary phase diagrams at its boundaries and generally is much complicated in understanding compared to binary phase diagrams. All the ternary compounds formed along with all the binary compounds among the components of the phase diagrams are important. Details of the different ternary and binary compounds formed in the SiO_2–CaO–Al_2O_3, SiO_2–CaO–FeO and SiO_2–CaO–Fe_2O_3 phase diagrams are provided in Tables 5.4, 5.5 and 5.6, respectively.

Any silica refractory contains very high amount of silica (generally above 90 wt%), and the major secondary oxide present is lime, added as a mineralizer. Hence, for silica refractories the primary phase present, even at the application temperature, is silica only, either tridymite or cristobalite, as per the prevailing temperature of application. However, in the near vicinity of the impurities or additives, minor amount of secondary phases may be present, depending on the constituents present in that zone. In SiO_2–CaO–Al_2O_3 (Figure 5.4) system, there are two ternary compounds formed, namely, anorthite and gehlenite, and, as per the phase diagram, liquid phase may exist even at temperatures below 1300°C. So liquid phase may be present in the close region of the impurity/additive particle even from a temperature <1300°C. The higher be the amount of the impurity phase, the greater will be the amount of liquid phase formed, and may affect the property and performance of the refractory. Hence, it is important to reduce the amount of impurity phase to minimize the amount of liquid phase formation and deterioration of properties. Similarly, in the SiO_2–CaO–FeO phase diagram (Figure 5.5), the liquid phase exists below 1100°C in the binary

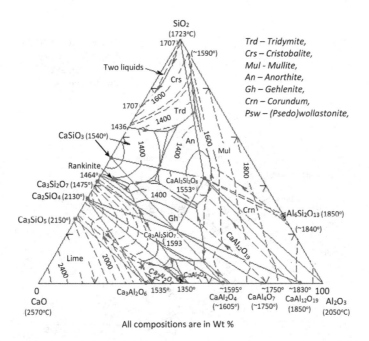

FIGURE 5.4 Phase diagram for the system SiO₂–CaO–Al₂O₃.

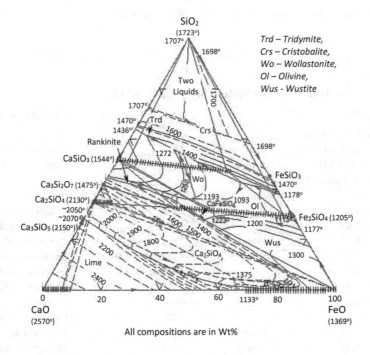

FIGURE 5.5 Phase diagram for the system SiO₂–CaO–FeO.

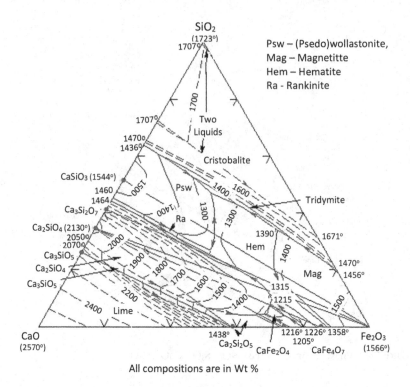

FIGURE 5.6   Phase diagram for the system SiO_2–CaO–Fe_2O_3.

TABLE 5.4

Compounds form in the System SiO_2–CaO–Al_2O_3

Formula	Name	Melting Point (°C)	Melting Character
Ternary Compound			
$2CaO.Al_2O_3.SiO_2$	Gehlenite	1590	Congruent melting
$CaO.Al_2O_3.2SiO_2$	Anorthite	1550	Congruent melting
Binary Compound			
$3CaO.SiO_2$	Alite	2150	Incongruent melting
$2CaO.SiO_2$	Balite	2130	Congruent melting
$3CaO.2SiO_2$	Rankinite	Dissociates at 1475	Incongruent melting
$CaO.SiO_2$	Wollastonite	1540	Congruent melting
$3Al_2O_3.2SiO_2$	Mullite	Dissociates at 1810	Incongruent melting
$3CaO.Al_2O_3$		Dissociates at 1535	Incongruent melting
$5CaO.3Al_2O_3$		1455	Congruent melting
$CaO.Al_2O_3$		1600	Congruent melting
$3CaO.5Al_2O_3$		1720	Congruent melting.
$CaO.2Al_2O_3$	Grossite	1789	Congruent melting
$CaO.6Al_2O_3$	Hibonite	1870	Incongruent melting

TABLE 5.5
Compounds form in the System SiO_2–CaO–FeO

Formula	Name	Melting Point (°C)	Melting Character
Ternary Compound			
$Fe_2Al_4Si_5O_{18}$	Iron cordierite	~1560	Incongruent melting
Binary Compound			
$3CaO.SiO_2$	Alite	2150	Incongruent melting
$2CaO.SiO_2$	Belite	2130	Congruent melting
$3CaO.2SiO_2$	Rankinite	Dissociates at 1475	Incongruent melting
$CaO.SiO_2$	Wollastonite	1540	Congruent melting
$FeO.SiO_2$	Ferrosilite	~1500	Incongruent melting
$FeO.2SiO_2$	fayalite	1205	Congruent melting

TABLE 5.6
Compounds form in the system SiO_2–CaO–Fe_2O_3

Formula	Name	Melting Point (°C)	Melting Character
Binary Compound			
$3CaO.SiO_2$	Alite	2150	Incongruent melting
$2CaO.SiO_2$	Belite	2130	Congruent melting
$3CaO.2SiO_2$	Rankinite	Dissociates at 1475	Incongruent melting
$CaO.SiO_2$	Wollastonite	1540	Congruent melting
$2CaO.Fe_2O_3$	Srebrodolskite	1450	Congruent melting
$CaO.Fe_2O_3$	Harmunite	1216	Incongruent melting
$CaO.2Fe_2O_3$		1228	Incongruent melting

part of CaO–FeO system and also for the ternary system, close to the ternary compound kirschteinite. Again, for SiO_2–CaO–Fe_2O_3 phase diagram (Figure 5.6), liquid phase exists from below 1200°C in CaO–Fe_2O_3 binary part and also in the ternary diagram. But the composition of the low-temperature liquids that are forming in the different phase diagrams are far away from the silica refractories. Hence, the chance of these liquid phase formation in actual application is very rare; and, even the liquid phase may form, it will be in very small amount and only in the close vicinity of the impurity particles. Thus, the effect of these liquid phase formation, if any, on the property and performance of silica refractory will be very limited.

5.7 MAIN APPLICATION AREAS

Silica refractories are mostly used where the atmosphere is acidic in nature. Great care is required for application of silica refractories, when it is heated from room temperature to at least 600°C due to the volumetric expansions associated with various polymorphic changes on increasing temperature. Allowance for expansion has to be provided for the whole lining, and loosening of structure (by tie rods) is done at different temperature ranges associated with the polymorphic changes.

There are three major application areas of silica refractories. The properties required for the silica refractories used for this application are detailed in Table 5.7. The main applications of silica refractories are:

    a. Crown/roof of the glass tank furnace
    b. Coke oven batteries
    c. Hot blast stove

5.7.1 CROWN/ROOF OF THE GLASS TANK FURNACE

Among the three major applications of silica refractories, this is the most critical one. Continuous glass manufacturing process involves the glass-melting tank furnace, and the melting chamber of the furnace has the most important function. Reaction among the batch materials occurs here with the melting of the glass. Figure 5.7a shows the internal view of a glass tank furnace melting zone with the crown or roof. The roof of this melting chamber (as detailed in Figure 5.7b) reaches above 1600°C and remains at such high temperature during the service life of the glass-melting tank. So, the refractories need to be very pure for such applications. Silica is preferred as the cristobalite is a very stable phase at that temperature range and also, if due to any reason, there is a chipping of refractory that will mix up with the glass batch (consisting about 75% of silica) without affecting the chemical composition and other properties of glass much.

TABLE 5.7

Details of the Properties Required for Silica Refractories That Are Used in Different Applications

Properties	Glass Tank Furnace Crown	Coke Oven Batteries	Hot Blast Stove
SiO_2 content (min) %	97	94	94
CaO content (max) %	2	2.5	2.5 – 3.0
$Al_2O_3 + TiO_2$ content (max) %	0.5	1.5	1.5
Fe_2O_3 content (max) %	0.5	1.5	1.0
Refractoriness, °C	1690	1680	1680
RUL (Ta), °C	1650–1680	1620	1620
Bulk density, g/cc	1.9	1.9	1.8 – 1.9
Apparent porosity (%)	18–20	18–20	24–26
CCS (Mpa)	35–40	30–35	30–35
Thermal expansion at 1000°C	1.26	1.18–1.2	1.22
PLCR, at 1450°C for 2 hours, % (max)	+ 1.0	+0.4	+ 0.4
Main-phase content	Cristobalite	Tridymite	Tridymite

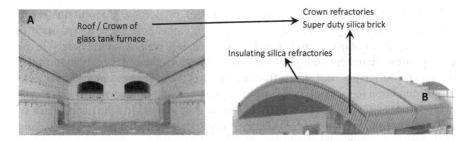

FIGURE 5.7 Roof or crown of the glass-melting tank furnace.

Also, with time, as the demand increases, the glass manufacturers are trying to increase the productivity and intensify the melting process. This demands for high quality of refractory linings, especially for crowns. The crown refractories are exposed mainly to alkali vapors (alkalis are part of glass batch material), which are very aggressive to acidic silica refractories. There are also chances of condensation of these alkali chemicals on the crown refractory surface and aggressive chemical reactions with the refractories. These chemical attacks intensify, especially when oxygen combustion process is used, and the temperature of the crown reaches up to about 1650°C and above.

But this application demands some special requirement of properties in silica refractories. Primarily, the refractories have to be very pure (SiO_2 ~96%–97%, CaO ~2.0%–2.5%). Impurities present in the refractory must be very minimum, as they affect the performance in the long run at high temperatures and may form liquid phase by reacting with the alkalis (vapor from batch materials) and silica. This may cause dripping of refractory and wear out of the crown. Also, these refractories need to be low in porosity, as porosities may allow the alkali vapor to enter within the interior of the refractory and form low-melting alkali silicates compounds. The conventional glass is a soda (Na_2O)–lime (CaO)–silica composition, and the alkali, Na_2O, may fly off with the flame and flue gas, reacts with silica and its mineralizer and impurities, and may reduce the liquidus temperature drastically. The even liquid phase may form at a temperature of 800°C in Na_2O–SiO_2 system, thus affecting the roof lining very badly. Formation of such liquid phases may cause deterioration and collapse of the crown structure. Refractories used in crown need to have high creep resistance, as they have to retain the structural integrity at very high temperatures for prolong time.

5.7.2 Coke Oven Batteries

Coke is the carbonization product of coal in reducing atmosphere done at about 1100°C after removal of volatile components. Coke is used mainly in the blast furnace for making iron, both as a reducing agent and as a source of thermal energy. At high temperature, it reduces the iron ore and helps to produce liquid metallic iron. Coal is not used directly in blast furnace due to its high volatile matter (VM) content, higher expansion on heating, crumbliness, and low strength. Coal is carbonized at high temperature up to a certain degree of volatilization to produce metallurgical coke of desired mechanical and thermochemical properties. Conventional coke making is done in a special type of heat chamber called coke oven; it consists of a series of a battery of ovens sandwiched between heating walls. Coal is charged in the coking (oven) chambers and gets heat from the heating chambers on both the sides through the refractory wall. The refractory lining of the coke oven coking chamber at cooled condition (open door) is shown in Figures 5.8a and during operational condition (closed) as Figure 5.8b, respectively. Figure 5.8c shows the schematic details of the coke oven. Current trend for coke ovens is to enlarge the dimensions of coking chambers to accommodate more amount of coal and enhance the coke production. This results in decreased lining thickness of the heat-exchanging parts of heating walls. Thus, lesser thick walls need to perform under greater productivity with high aggressive conditions.

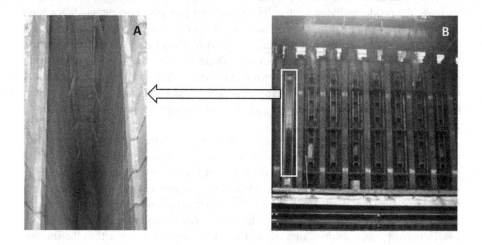

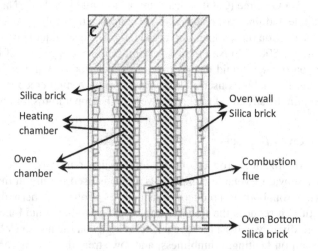

FIGURE 5.8 Coke oven batteries for iron and steel industries.

Silica refractory is commonly used in the construction of a coke oven battery and especially with tridymite phase. The main reason for selecting silica refractory for coke ovens are: (i) tridymite phase of silica refractory has the thermodynamic stability in the temperature range up to 1470°C (generally coke ovens operate between 800°C and 1350°C), and (ii) it has a very minimal creep rate at that operating temperatures. Also, tridymite has the minimum thermal expansion properties among the different crystallographic forms of silica (Figure 5.2), so it results in better thermal shock resistance during application. Almost all the expansions of silica brick take place below the coke oven operating temperatures. So during normal operation of the coke oven, fluctuations in temperature due to the opening of the battery, during charging of coal and discharging of coke, do not affect much on the volume stability of the refractory wall. Again charging and discharging involve friction and abrasion of hard, angular, sharp edged coal, and coke particles with the refractory wall, causing wear and abrasion of the refractory wall and the floor. Hence high abrasion resistance, high strength, and low porosity are required for the refractories. The addition of zircon is done to coke oven silica refractories to improve the abrasion and wear resistances. Also, the heating of coal in the coking chamber is done by passing heat through the wall of the heating chambers placed on both the sides. Hence, high thermal conductivity with excellent creep resistance at the operating temperatures is required for the refractories. The addition of silicon carbide is also done to improve the thermal conductivity of the refractories. There are many critical shapes and sizes of silica refractories required for coke oven applications, and these are prepared by hand-molding techniques. These refractories also need to satisfy the properties as mentioned.

5.7.3 HOT BLAST STOVE

It is a unit used by the metallurgical industries, especially for iron making, which preheats the air blown into the blast furnace by utilizing the waste heat of the hot flue gases. The stove is a tall, cylindrical steel shell covered unit lined with refractory. The inner portion is separated into two chambers: a combustion chamber, in which gases from the blast furnace and other fuel sources such as the coke ovens are burnt in the presence of combustion air, and a regenerative chamber lined with a checker work of refractory bricks heated by the burnt gas. More than one stove is required by the industries; when one stove is getting heated by the hot flue gas, the cold air blast passes through the other stove (already preheated) and gets heated by the heat stored in the regenerative chamber on its way to the blast furnace. In this way, the air is preheated up to the maximum of 1300°C before entering the blast furnace. A schematic detail of the hot blast stove is shown in Figure 5.9 with the checker silica refractory in the inset. In this figure, solid arrows mark the direction movement of hot waste gas, and dotted arrows show the movement of cold air that is getting heated and coming out as hot preheated air.

In the stoves, silica refractories are used in many areas, namely, walls, domes, and the checker work. The main properties required for this applications are high volume stability, high creep resistance, and thermal shock resistance. As the temperature of stove varies between 800°C and 1300°C, tridymite phase is the most stable form of silica refractory, which is volumetrically stable and has high creep resistance. Also,

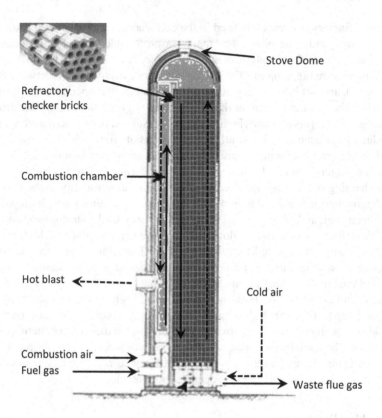

FIGURE 5.9   Schematic of hot blast stove and silica bricks/shapes (inset).

high thermal conductivity with low residual quartz content (<1%) and low expansion properties are desirable for this application.

5.8   SILICOSIS

It is also important to know about any harmful effect of the refractories to our biological system. Working for a long time in silica plants where a human is frequently exposed to high silica-containing dust, whether produced by cutting and grinding of silica refractory or by quartzite powder, is highly dangerous, as it may cause silicosis. As per medical term, silicosis is the fibrosis (swelling) of lungs due to inhalation of dust containing silica. Initially, it was believed that silicosis is a purely physical activity where the lungs are lacerated by the sharp and tiny hard grains of silica. Later, it was proved that not only physical but also chemical activity is involved for the disease due to acidic character of silica.

Silicosis occurs due to deposition of fine respirable dust (less than 10 μm in diameter) containing crystalline silica in the form of alpha-quartz, cristobalite, or tridymite. The disease is characterized by shortness of breath, cough, fever, and cyanosis (bluish skin). It may often be misdiagnosed as pulmonary edema (fluid in the lungs), pneumonia, or tuberculosis.

SUMMARY OF THE CHAPTER

Silica is one of the most abundant minerals available in nature as pure form and is used for making silica refractory.

Silica has three polymorphic forms, namely, quartz, tridymite, and cristobalite, and each form has low- and high-temperature variations. Each polymorphic form and their temperature-dependent variations transform from one to another on changing the temperature but is associated with volume change. These changes are crucial for silica manufacturing and their applications, as these may result in cracking of the refractory.

Polymorphic transformations are sluggish in nature, and they do not complete during firing. To accelerate them, special chemicals called mineralizers are added in silica-refractory batch during mixing. To accelerate and stabilize cristobalite and tridymite phases, CaO and mix of $CaO + Fe_2O_3$ are added, respectively.

The firing of silica refractories needs to be very slow to accommodate the volume changes without structural disturbances and to crack. The final temperature is dependent on the desired phase in the fired refractory.

The presence of free quartz in fired refractories called residual quartz (RQ) is highly detrimental, as it converts to high-temperature forms during firing and use, causing cracks and failure of the refractory.

Properties of the silica refractories are dependent on the purity, and amount and types of impurities present. In general, silica refractory has an RUL value very close to its refractoriness, due to its purity and, in-case any liquid phase is formed from impurities, the viscosity of the liquid remains very high due to high silica content, and deformation of the shape is very less.

Silica refractories are mostly applied to the crown of glass-melting tank furnace, coke oven batteries, and hot blast stoves in iron and steel industries.

QUESTIONS AND ASSIGNMENTS

1. Discuss the different crystallographic forms of silica.
2. Why a manufacturer or a user of silica refractory needs to be careful during heating or cooling of the refractory?
3. Describe in detail about the manufacturing technique of silica refractory.
4. Write the precautions during firing of silica refractory.
5. What is a mineralizer? What mineralizers are used in silica refractory?
6. What is the Dales theory/How CaO works in stabilizing silica refractory?
7. Why silica refractories have RUL values very close to that of refractoriness?
8. Describe the effect of impurities present in silica refractories using the phase diagrams.
9. Describe different applications of silica refractory.
10. Discuss the specific properties required for silica refractories in each of the three main applications.
11. What is residual quartz in silica refractories, and how it is important?
12. Why is alumina a harmful impurity in silica refractories?
13. What do you know about silicosis?

BIBLIOGRAPHY

1. J. H. Chesters, *Refractories- Production and Properties*, Woodhead Publishing Ltd, Cambridge, 2006.
2. C. A. Schacht, *Refractories Handbook*, CRC Press, Boca Raton, US, 2004.
3. P. P. Budnikov, *The Technology of Ceramics and Refractories*, Translated by Scripta Technica, Edward Arnold, The MIT Press, Massachusetts, US, 4th Ed., 2003.
4. A. R. Chesti, *Refractories: Manufacture, Properties and Applications*, Prentice-Hall of India, New Delhi, India, 1986.
5. *Refractories Handbook*, The Technical Association of Refractories Japan, Tokyo, 1998.
6. *Harbison Walker Handbook of Refractory Practice*, Harbison-Walker Refractories Company, Hassel Street Press, Moon Township, PA, 2005.
7. F. Brunk, Silica refractories, CN refractories, *Special Issue*, 5 27–30 (2001).
8. F. Brunk, Silica bricks for modern coke oven batteries, *Coke making International*, 2 37–40 (2000).
9. I. A. Aksay and J. A. Pask, Stable and metastable equilibria in the system SiO_2-Al_2O_3, *Journal of the American Ceramic Society*, 58 [11–12] 507–512 (1975).
10. J.R. Taylor and A.T. Dinsdale, Thermodynamic and phase diagram data for the CaO-SiO_2 system, *Calphad*, 14 [1] 71–88 (1990).
11. P. Wu, G. Eriksson, A. D. Pelton and M. Blander, Prediction of t he thermodynamic properties and phase diagrams of silicate systems – evaluation of FeO – MgO – SiO_2 system, *ISIJ International*, 33 [1] 26–35 (1993).
12. R. C. DeVries, R. Roy and E. F. Osborn, The system $TiO2$-SiO_2, *Transactions of the British Ceramic Society*, 53 [9] 525–540 (1954).
13. M. Kotouček, P. Kovář, K. Lang, L. Vašica and L. Nevřivová, Dense silica – Properties, production and perspectives. *Refractories Worldforum*, 5 [2] 65–68 (2013).
14. E. F. Osborn, and A. Muan, The system CaO–Al_2O_3–SiO_2. In *Plate No 1 in Phase Equilibrium Diagrams of Oxide Systems*. The American Ceramic Society, Columbus, OH, 1960, pp. 219–221.
15. E. F. Osborn and A. Muan, The system CaO-FeO-SiO_2. In *Plate 7 in Phase Equilibrium Diagrams of Oxide Systems*. The American Ceramic Society, Columbus, OH, 1960.
16. E. F. Osborn and A. Muan, System CaO-Fe_2O_3-SiO_2. In *Plate 10 in Phase Equilibrium Diagrams of Oxide Systems*. The American Ceramic Society, Columbus, OH, 1960.
17. M. Kotouček, P. Kovář, K. Lang, L. Vašica and L. Nevřivová, Dense silica – Properties, production and perspectives. *Refractories World Forum*, 5 [2] 65–68 (2013).

6 Alumina Refractories

6.1 INTRODUCTION

Alumina (Al_2O_3) refractories cover a wide range of refractories, are most common and have the major portion of the total refractories. Any refractory containing more than 50 wt% Al_2O_3 is termed as alumina refractory. Low-alumina-containing (say 50%) refractories are a better version of fire clay refractories having greater volume stability, higher strength both at ambient and elevated temperatures, improved resistances against corrosive attack, and better resistances against abrasion and erosion, thermal shock, and creep compared to other conventional refractories. Again, higher amount of Al_2O_3-containing refractories show significantly different and improved properties than that of the low Al_2O_3-containing ones.

During the first-half of the 1940s, major development of Al_2O_3 refractories took place when the use of Al_2O_3–SiO_2 bricks, using bauxite as major raw material, started in many of the industrial applications. The commercial success of the Bayer process for manufacturing very high pure alumina (and aluminum) opened up a new horizon for the manufacturing of high pure-alumina refractories. This process also gave an entirely new class of alumina raw materials like calcined, fused, and tabular alumina, generating a new line of high Al_2O_3 refractories. Also, development in processing technology like grinding, separation, mixing, pressing, and firing have improved considerably with time and resulted in a technological advancement in the refractory manufacturing. Technical improvements in all types of refractory products have resulted in better performance and reduced consumption. The result has been observed as a decline in production of conventional refractories, especially the fire clay-based Al_2O_3–SiO_2 refractories.

Alumina refractories are having silica as the second most prominent oxide phase, and they cover the major portion of the Al_2O_3–SiO_2 phase equilibrium diagram, as shown in Figure 6.1. Low-alumina- and high-silica-containing regions of the phase diagram are representing the silica and fireclay refractories. If we increase the alumina content in pure silica composition (starting from zero level of alumina), the refractories that appear first are super duty silica refractory (~97% SiO_2 and ~0.5% Al_2O_3). A little increase in alumina content and reduction in silica produces the general purpose silica refractory (~94% SiO_2 and ~2% Al_2O_3). Increase in alumina produces the semi-silica fire clay refractory (65%–80% SiO_2 and 18%–25% Al_2O_3). Further increase in alumina content produces the fireclay refractories containing ~25–45 wt% Al_2O_3 and then comes the alumina refractories. Again, the alumina refractories are classified as per the Al_2O_3 content in it. Generally, the classifications are 50%, 60%, 70%, 80%, 90%, and 99% Al_2O_3-containing refractories. All the major classes of refractories based on the Al_2O_3–SiO_2 phase diagram are marked on Figure 6.1 for better understanding.

DOI: 10.1201/9781003227854-6

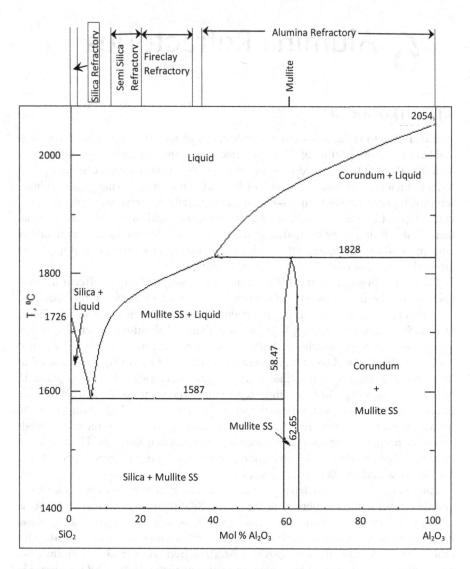

FIGURE 6.1 Phase diagram of alumina–silica system.

It can be seen from the phase diagram that the refractoriness of the refractory increases with the increase in the Al_2O_3 content. The only eutectic present is at 1587°C, has a composition of 94.5% SiO_2 and 5.5% Al_2O_3, and is far away from the alumina refractories. The only compound—mullite (3 Al_2O_3 2 SiO_2) forms at a composition of 71.8 wt% Al_2O_3 and 28.2 wt% SiO_2. There are different opinions regarding the melting behavior of mullite; as per some researcher, it is a congruent melting compound having a sharp melting point at 1840°C, and, in contradiction, some researchers found it as an incongruently melting compound with decomposition and liquid formation at 1828°C. As on toady, mullite is being considered as an incongruently melting compound with a defective lattice structure that allows making some solid solution on both the sides of the composition, as described by Aksay and Pask, and the Figure 6.1 represents their work.

6.2 RAW MATERIALS AND SOURCES

Aluminum is the third most abundant element in the earth's crust, and so plenty of aluminum-bearing minerals are available in nature. But nearly no naturally occurring commercial source is available in pure oxide form (alumina). Most commonly available minerals containing a high amount of alumina with the minimum impurities are used as raw materials for alumina refractories. As the alumina content varies widely in alumina refractories (say from 50% to 99%), different types of natural (and few synthetic) raw materials are used, having different amounts of alumina, to obtain the required alumina content in the final composition with the economy. Synthetic raw materials are used for making very high pure-alumina refractories, of which purity level is difficult to obtain from natural sources and also to impart some special properties in the refractories. The commonly used raw materials for alumina are as follows.

6.2.1 Fireclay

Details of the fire clay are described in Chapter 7, dealing with fireclay refractories. For making alumina refractories with low alumina content (50%–60% Al_2O_3) calcined fire clays are used in combination with some high-alumina clays (commonly termed as diaspore clay or bauxitic clays) or natural anhydrous alumino silicate minerals. A good-quality calcined fire clay contains 40%–45% Al_2O_3 and to get the balanced amount of alumina the batch composition is adjusted with the higher alumina-containing raw materials. As the fireclay contains water in its structure, it has to be calcined before use in refractory composition.

6.2.2 Anhydrous Aluminosilicates

Naturally occurring anhydrous aluminosilicates, namely, sillimanite, kyanite, and andalusite are having a same chemical formula, Al_2SiO_5, but differing crystal structures. The differences in crystal structure give somewhat unique physical properties to each of the three minerals, and each one is used in slightly different ways for making alumina refractories. In theoretically pure form, the aluminosilicates contain 62.9% Al_2O_3 and 37.1% SiO_2, but the natural resources are impure, and alumina content is less than 60%. Main impurities present are Fe_2O_3, TiO_2 and CaO. Sillimanite is not widely available and used in lesser extent due to limited availability in rock form. Some beach sand sources of sillimanite are there in very few countries. Comparatively, much greater availability and widely used mineral is kyanite, but mostly it is contaminated with quartz. Andalusite is also widely available throughout the globe, but many sources are associated with pyrophyllite clay mass as an impurity.

TABLE 6.1
Properties of Sillimanite, Kyanite, and Andalusite

	Sillimanite	Kyanite	Andalusite
Crystal system	Orthorhombic	Triclinic	Orthorhombic
Specific gravity	3.23	3.56–3.66	3.16–3.20
Hardness, Moh's scale	6–7	5–7	7.5
Mullite formation temperature	1550°C–1650°C	1100°C–1480°C	1450°C–1500°C
Volume changes on calcination	Slight increase	Very large increase	Very slight increase
Specific gravity after calcination	3.10	3.05	3.04
Fusion temp	>1800°C	>1800°C	>1800°C

These aluminosilicate compounds are not present in the very common Al_2O_3–SiO_2 phase diagram that shows the equilibrium at 1 atm pressure. None of these three minerals are equilibrium phases at 1 atm. These are formed at high-pressure and high-temperature geological conditions in the earth, and all are available as a metastable phase in nature. On heating, these metastable aluminosilicates convert to equilibrium-stable mullite phase and free silica glass. The temperature of this decomposition and mullite formation changes from mineral to mineral. Table 6.1 summarizes the properties of these three natural aluminosilicates and the differences among them. Kyanite has the highest specific gravity among the three alumina silicates, and on heating converts to mullite and silica glass having significantly lower specific gravity. Such conversion results in a huge volume expansion. Hence, kyanite has to be calcined before use in alumina refractory.

6.2.3 BAUXITE

Bauxite is the most commonly available and widely used aluminum ore, which consists mostly of the minerals gibbsite [$Al(OH)_3$], boehmite [γ-$AlO(OH)$] and diaspore [α-$AlO(OH)$]. Truly speaking, bauxite is not a mineral but rather a mix of different aluminum hydroxides. The term is used to describe the economically important mixture of these minerals, which forms a mass of the individually classified members of gibbsite, boehmite and diaspore. All the bauxites that are available in nature are not easily extractable from the main ore/rock, and separation of the aluminum-bearing mineral is sometimes difficult or costly affairs. The bauxite sources that require less-complicated process to separate out the aluminum-containing phases are called as recoverable bauxites, and those sources and reserves are economically important. Again, the major application of bauxite is for manufacturing of aluminum metal that involves a chemical treatment of raw bauxite first to separate out the only-aluminum phase. Thus, purity of this metallurgical grade bauxite may be little compromised, as chemical reactions involved in the processing removes the impurities from the system easily. But for the use in refractory industries, the bauxite is only calcined without any chemical treatment. Hence, the refractory-grade bauxites need to purer.

Theoretically, bauxite contains 73.9% Al_2O_3 and 26.1% H_2O, and, on the calcined basis, it is 100% Al_2O_3. In natural occurrence, mixtures of gibbsite and boehmite are common in bauxite and mix of boehmite, and diaspore is less common and gibbsite and diaspore are rare. Some high-grade refractory bauxites consist solely of gibbsite with minor amounts of kaolin clay [$Al_2Si_2O_5(OH)_4$]. The common impurities found in both metallurgical- and refractory-grade bauxites are aluminosilicate (mainly clay), quartz (SiO_2), hematite (Fe_2O_3), goethite [$FeO(OH)$], rutile (TiO_2) and anatase (TiO_2). Hematite and goethite are the most abundant iron impurities in many bauxites and the principal reason for the red and brown colors that are characteristic of countless bauxite deposits.

World bauxite reserves are estimated to be 55–75 billion tons. But all the reserve is not viable for commercial exploitation, and the recoverable reserves are estimated to be around 25 billion tons. The largest recoverable reserves are in Australia, followed by Guinea, Brazil, Jamaica, and India. Among different nonmetallurgical grades of bauxite, the refractory grade is the purest one. A bauxite to be used in refractories should have at least 58% of Al_2O_3 content, and the impurities that may be allowed are a maximum of 5% of SiO_2, the maximum of ~3% of TiO_2 and the maximum of 3% of Fe_2O_3. Since bauxites give away their water during heating accompanied by a marked reduction in volume, they are calcined and sintered before used in refractories. During this sintering, corundum and mullite phases are formed, together with a small amount of low-melting liquid phase containing iron and titanium.

6.2.4 SYNTHETIC RAW MATERIALS

Naturally occurring mineral, mainly bauxite, is purified by chemical treatment to remove the impurities and is further calcined to produce synthetic alumina. Chemical treatment process of bauxite to produce pure alumina to be further processed to manufacture aluminum metal was first developed by C. J. Bayer in 1888, which is still being used industrially, and commercially without much modification is termed as Bayers' process. Synthetic alumina, as obtained from the Bayers' process as an intermediate product in metallic aluminum manufacturing, is available mainly in three forms: activated alumina, smelter-grade alumina, and calcined alumina. The porous, granular, activated alumina aggressively absorbs liquid water and water vapor. Smelter-grade alumina is used for making aluminum metal. The fine-grain calcined alumina is a dense, impermeable ceramic material used for abrasives, refractories, electrical insulation, high-temperature crucibles, and dental restoration. It is also used as a filler for paints, glass, and ceramics.

Fused alumina is also used for making alumina refractories for special properties, mainly to improve the corrosion resistance. They are produced by electrofusion route. In fused alumina, alumina crystals grow from a molten stage, and the growth rate is very high compared to solid-state growth resulting in very large crystals when solidified as fused alumina. Large crystals have lesser numbers of grain and grain boundaries, resulting in lesser surface area for reaction by any corrosive agent and improves the corrosion resistance. Fused alumina can be prepared from the bauxite directly, having impurities like iron oxide, silica, and titania with a reddish/yellowish/brownish color called brown, fused alumina. Again, a white fused alumina with the minimum impurities can also be obtained by fusing Bayer's process purer alumina.

Sintered alumina is another variety of synthetic alumina prepared from finely ground calcined Al_2O_3 by sintering it below its melting point. As a result of this high-temperature sintering process, the properties are excellent, and, in particular, a uniform crystal structure with high strength is attained both at ambient and elevated temperatures.

There is also a special class of sintered alumina available commercially, called tabular alumina, in which shaping of high, pure calcined alumina is done in a special technique, and sintering is done at very high temperature, so that the processing enables the alpha alumina crystals to grow into large grains, resembling like tablets, and hence termed as "tabular" alumina. In this processing, the calcined alumina particles are shaped into spheres for firing that shrink and result in a uniform microstructure with low residual porosity. The sintered alumina spheres are crushed and ground to different size fractions. Tabular alumina has excellent properties like high density, low open porosity, dimensional stability, creep and abrasion resistance, exceptional resistance to thermal shock, and, uniform, compact microstructure.

Synthetic mullite ($Al_6Si_2O_{13}$) is also used as raw material for some special alumina refractories, produced by the sintering process or fusion. In the process, the initial materials clay or kaolin are enriched in alumina content to have a composition of mullite by the addition of calcined Al_2O_3. The mixture is then electrofused or pressed and sintered as per the processing, and the fired products are crushed and ground to get the desired particles' fractions.

6.3   BRIEF OF MANUFACTURING TECHNIQUES

Alumina refractories cover a wide class of refractories having a wide variety of alumina content and other secondary oxides. Accordingly, the raw materials also vary according to the composition of the refractory to optimize the properties and economy. Table 6.2 provides the classification of alumina refractory and the main raw materials used for making them.

Alumina refractories are manufactured by the typical manufacturing process of refractory making. Different fractions of the raw materials are taken and mixed in

TABLE 6.2
Raw Materials Combination for Different Alumina Refractories

Refractory Class	Raw Materials Used
50% Al_2O_3	Fireclay, diasporic clay, aluminosilicates, bauxite
60% Al_2O_3	Aluminosilicate, bauxite, fire clay, bauxite clay
70% Al_2O_3	Bauxite, aluminosilicates, bauxite clay
80% Al_2O_3	Bauxite
90% Al_2O_3	Bauxite, synthetic alumina
> 95% Al_2O_3	Synthetic alumina

pan/muller/counter-current mixer, initially as dried condition, and then green binders and moisture are added. Binders provide proper binding and handling strength at the green and dried conditions. Mixed batch is then shaped by pressing, and then they are dried and fired. Pressing of alumina refractories is done in mechanical, friction screw, or hydraulic presses. Binders and moisture also impart some plastic character to the mix to retain the shape after pressing/shaping. Firing is done in both the batches or continuous type of kilns, and the temperature is dependent on the composition, especially on the alumina content, and may vary between 1450°C and 1750°C.

6.4 CLASSIFICATIONS AND PROPERTIES

Classification of alumina refractories is done as per the alumina content, and obviously the properties will also vary. Table 6.3 shows a general idea about the properties of the different alumina refractories, including a general chemical composition of each category. In general, the second component in the alumina refractories is silica, which reacts with alumina and forms mullite. Mullite, having a high volume stability, lower thermal expansion, high creep resistance, high thermal shock resistance, is beneficial for alumina refractories. But the presence of impurities like iron oxide, titania, lime and alkalis may react with alumina and alumio silicate phases, and form low-melting compounds, and thus strongly deteriorate the high-temperature properties. An increasing amount of alumina content produces a better refractory with higher strength, density, and hot properties. The presence of higher amount of iron oxide may reduce the hot strength (say, RUL) drastically, even the composition may have a higher alumina content. Also, alumina shows little acidic character at high temperatures and may form compounds on reaction with highly basic materials.

However, as the same quality of alumina refractory can be prepared by a different source of raw materials, the properties may again vary as per the raw materials

TABLE 6.3
Properties of Different Alumina Refractories

Refractory Class	Al_2O_3%	Fe_2O_3%	BD (g/cc)	AP (%)	CCS (MPa)	RUL, Ta (°C)
50% Al_2O_3	50	1.3	2.35	18	35	1500
60% Al_2O_3	60	3	2.4	22	40	1450
62% Al_2O_3	62	1.2	2.5	16	60	1550
70% Al_2O_3	70	2.5	2.6	20	40	1480
70% Al_2O_3 (mullite)	70	0.8	2.55	18	60	1680
80% Al_2O_3	80	2.5	2.7	22	50	1500
85% Al_2O_3	85	1.5	2.9	18	60	1600
90% Al_2O_3	88 min	0.5	2.95	18	65	1700
95% Al_2O_3	94 min	0.5	3.0	22	70	1700
99% Al_2O_3	97 min	0.1	~3.1	18	75	1750

used. Accordingly, the applications are to be selected. Otherwise, the same amount of alumina-containing refractory may perform differently. For example, 50% and 60% Al_2O_3-containing refractories, when prepared from bauxite or andalusite, typically exhibit high reheat expansion while refractories based on clay (fire clay, diasporic/bauxite clay) do not. Thus, there is a fundamental difference in the refractories within the same class (same $Al_2O_3\%$) on permanent expansion characteristics. In linings requiring the extreme tightness (as in rotary kiln applications), the reheat expansion is important and improves the lining life. But, in contrast, high reheat expansion may be associated with the high spalling tendency, i.e., low thermal shock resistance. In this regard, refractories produced from a clay base material have superior properties. This is because of their finer texture, smaller average pore size, and due to the absence of permanent expansion on heating. Similar kind of character is also evident for the 70% Al_2O_3-containing refractories.

Again, the presence of higher amount of impurities, say iron oxide, will reduce the liquidus temperature by forming low-melting compounds on reaction with alumina and silica (impurity), and will result in lower hot strength (RUL) properties. Thus, the same class of refractory may have different properties and will have different applications.

Refractories containing Al_2O_3 in the range of 80%– 85% were originally developed for use in aluminum smelting and holding furnaces. These refractories are purely based on calcined bauxite, as it is the closest mineral in Al_2O_3 content to their overall composition. The resistance against molten aluminum and salt fluxes for these refractories is coming from the resistance of the bauxite. But these refractories are not highly successful in the ferrous industry. The reason for that is the processing temperature. The aluminum industry operates at a much lower temperature than the ferrous industries. At higher temperatures the bond phase (glass and mullite) of the refractories, holding the bauxite aggregates, gets softened. Hence, in an aggressive slagging situation at high temperatures, the bauxite aggregates are eroded out of the refractory brick due to soft and weak bond phase, and the wear rates are unusually high.

Refractories containing Al_2O_3 in the range of 90% or above are having the highest strength and erosion, and corrosion resistance and hot properties. These refractories are mainly made up of synthetic Al_2O_3 aggregates, and contain fused Al_2O_3 for improved corrosion and erosion resistances in some special cases. Depending upon the impurity phase and its quantity, the bond phase may be mullite-based or alumina-based ones. Alumina-bonded alumina refractories (direct bond) result in superior properties, especially the corrosion and hot strength properties.

There is a practical limit on Al_2O_3 content in high-alumina refractories [~96% Al_2O_3 (contains ~3% SiO_2)] in refractory brick for the highest-temperature applications. Products containing further higher Al_2O_3 content are difficult to sinter in conventional firing techniques and industrially available temperatures (~1750°C). These refractories show comparatively poor properties with respect to their alumina content, especially for density, strength, and reheat change (PLCR) characteristics. Though, up to 99% Al_2O_3 containing refractories exist commercially, they are primarily used for low-temperature applications mainly for their inertness (chemical resistance).

6.5 EFFECT OF IMPURITIES WITH BINARY AND TERNARY PHASE DIAGRAMS

Impurities are also very effective in affecting the properties of the alumina refractories. The most common impurities present with alumina are silica, iron oxide, titania, and alkalis. In most of the alumina refractories, silica is the second major component, and presence of only silica does not deteriorate the properties of alumina very strongly. As seen in Figure 6.1, the presence of silica in high-alumina compositions will produce a mixture of corundum and mullite solid solution, and a liquid phase appears only at 1828°C. Above which the phase composition changes to corundum and liquid. Composition and amount of the liquid changes with the amount of silica present in the system and the temperature. Hence, presence of only silica as impurity will reduce the liquid-formation temperature up to 1828°C, which is much above the general application temperature of the alumina refractories. So deterioration in high-temperature properties does not affect the performance.

The presence of iron oxide also affects the liquidus and reduces the liquid-formation temperature, but the effect is also not so pronounced. For Fe_2O_3, alumina forms a corundum solid solution at high temperature, and thus absorbs the ferric ions in the corundum structure, and so the properties are hardly affected, as shown in Figure 6.2a. Again, the liquid phase starts forming from 1689°C at a higher level of Fe_2O_3, and a mixture of corundum solid solution with liquid phase was obtained above this temperature. For FeO (Figure 6.2b), the liquid phase starts forming at 1750°C, which is the eutectic temperature in the system. Also, a congruent melting spinel compound forms in the system, named hercynite ($FeAl_2O_4$, containing about 58% Al_2O_3) with a melting point of 1820°C. Hence, presence of only iron oxide as an impurity in alumina system, as a ferric or ferrous state, does not affect the hot properties strongly.

The presence of titania also shows (Figure 6.2c) similar characteristics as that of FeO, forming a eutectic and also forms a congruent melting compound—aluminum titanate. Eutectic is at 1843°C, and the compound melts at 1854°C. Hence, only TiO_2 is also not very effective in deteriorating the properties of alumina. But, the presence of lime (CaO) in alumina is effective in reducing the liquidus temperature (Figure 6.2d). The temperature decreases with the increasing amount of lime content, and a number of incongruent melting compounds also form in the system. The alumina-rich compound is calcium hexa aluminate (CA_6), which incongruently melts at 1850°C. Further increase in CaO produces calcium di aluminate (CA_2), which decomposes to liquid and CA_6 at 1760°C. On further increase in CaO, calcium aluminate forms, which incongruently melts to CA_2 and liquid at 1606°C. The lowest temperature of liquid formation in the $CaO–Al_2O_3$ system is at 1350°C, which is a eutectic of the system, having only about 36% alumina content. Though this composition never reaches in alumina refractories, at the site of the CaO impurity, there may be an initiation of the liquid phase at the surroundings of the lime particles present. Hence, that may affect the whole system. This temperature is quite low compared to any refractory application temperatures, and so the presence of lime in alumina needs to be carefully checked.

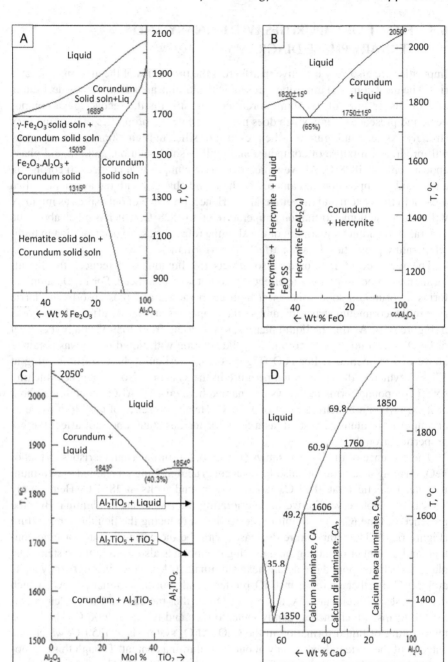

FIGURE 6.2   Phase diagrams of alumina with different impurity oxides, (a) Fe_2O_3, (b) FeO, (c) TiO_2 and (d) CaO.

The binary phase diagrams are useful for understanding the behavior of the refractory when only one single impurity phase is present. However, as refractories are prepared from natural materials, they contain a number of impurity ions and further are in contact with large number of ions present in application environment. So, compound effects of all the different types of impurities are prevailed on the refractories. Study of binary phase diagrams to understand and predict the behavior of refractories in real cases is far away from reality. Ternary phase diagrams are still closer to actual system, as it considers two separate impurity ions along with the main component. As alumina refractories mainly contain silica, iron oxide, titania, and lime as the major impurities, the ternary phase diagrams of Al_2O_3–CaO–SiO_2, Al_2O_3–SiO_2–FeO, Al_2O_3–SiO_2–TiO_2 and Al_2O_3–Fe_2O_3–TiO_2 are of prime importance.

Al_2O_3–CaO–SiO_2 ternary system is described in Chapter 5, with the phase diagram in Figure 5.4. For alumina refractories, the same phase diagram is important, but the alumina-rich compositions are of prime consideration. In this phase system, composition of the low-melting liquid phase moves away from the alumina-rich portion with decreasing temperature. The ternary compounds and the ternary eutectics present in the system are far away from the primary phase region of alumina. This indicates that commercially used alumina refractories that contain impurities coming from the raw materials are free from such liquid phase formation at their application temperatures.

Details of the other ternary phase diagrams, namely, Al_2O_3–SiO_2–FeO, Al_2O_3–SiO_2–TiO_2 and Al_2O_3–Fe_2O_3–TiO_2 are shown in Figures 6.3, 6.4 and 6.5, respectively,

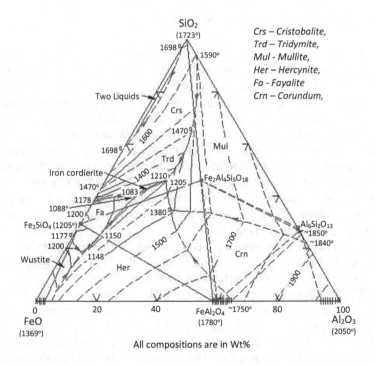

FIGURE 6.3 Phase diagram for Al_2O_3–SiO_2–FeO system.

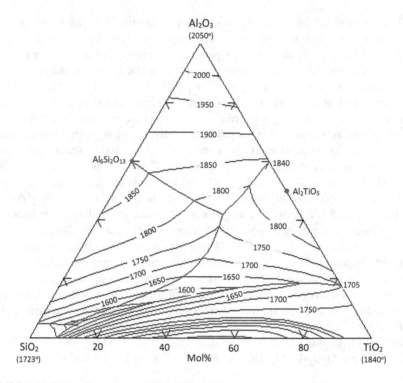

FIGURE 6.4 Phase diagram for Al₂O₃–SiO₂–TiO₂ system.

and the different ternary and binary compounds formed in these systems are compiled in Tables 6.4, 6.5 and 6.6, respectively. In Al₂O₃–SiO₂–FeO system (Figure 6.3), the ternary liquidus phase moves away from the alumina-rich portion, indicating that liquid phase exists at lower temperature away from the primary phase region of alumina. The lowest temperature for existence of any liquid phase in primary phase region of alumina is considerably higher than the application temperatures of alumina refractories. The ternary liquidus start from the binary invariant (eutectic/peritectic) points (composition) and move toward further lower temperature with the increase in the other components present. Iron Cordierite (2FeO·2Al₂O₃·5SiO₂) is the only compound formed in the system containing <40% of alumina and does not present in any commercial alumina refractories due to compositional mismatch. Thus, the liquid phase in the ternary system is not affecting the properties and performance of the alumina refractory. However, zonal effect of the low-melting phase formation in the close region of the impurity phase/particle present in the refractory may be observed, causing some impact on the properties of the whole refractory.

Ternary phase diagram containing Al₂O₃–SiO₂–TiO₂ (Figure 6.4) does not have any ternary compound. Two binary compounds are present among Al₂O₃–SiO₂ and Al₂O₃–TiO₂ systems, details are mentioned in Table 6.5. SiO₂–TiO₂ system (Figure 5.3c) does not have any binary compound but has a eutectic close to silica end with presence of liquid phase up to 1550°C. In the ternary system, the ternary liquidus moves close to the silica region with reducing temperature, and the liquid phase

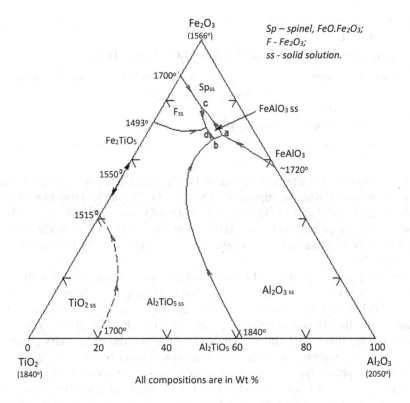

FIGURE 6.5 Phase diagram for Al_2O_3–Fe_2O_3–TiO_2 system.

TABLE 6.4

Compounds Formed in the System Al_2O_3–SiO_2–FeO

Formula	Name	Melting Point (°C)	Melting Character
	Ternary Compound		
$2FeO \cdot 2Al_2O_3 \cdot 5SiO_2$	Iron Cordierite (sekaninaite)	~1560	Incongruent melting
	Binary Compound		
$3Al_2O_3 \cdot 2SiO_2$	Mullite	Dissociates at 1810	Incongruent melting
$FeO \cdot Al_2O_3$	Hercynite	~1780	Congruent melting
$2FeO \cdot SiO_2$	Fayalite	1205	

TABLE 6.5

Compounds Formed in the System Al_2O_3–SiO_2–TiO_2

Formula	Name	Melting Point (°C)	Melting Character
	Binary Compound		
$3Al_2O_3 \cdot 2SiO_2$	Mullite	Dissociates at 1810	Incongruent melting
$Al_2O_3 \cdot TiO_2$	Aluminum titanate	1854	Congruent melting

TABLE 6.6

Compounds Formed in the System $Al_2O_3-Fe_2O_3-TiO_2$

Formula	Name	Melting Point (°C)	Melting Character
	Binary Compound		
$Al_2O_3 \cdot TiO_2$	Aluminum titanate	1854	Congruent melting
$Fe_2O_3 \cdot TiO_2$	Iron titanate (pseudobrookite)	1395	

may remain even up to a ternary eutectic at 1470°C. The ternary system has a peritectic point at 1480°C with a composition of SiO_2:Al_2O_3:TiO_2 molar ratio 84.87:5.1:9.97 and a eutectic point at 1470°C with SiO_2:Al_2O_3:TiO_2 molar ratio 84.43:4.72:10.85. Both the ternary eutectic and peritectic compositions are very close to silica end, far away from the alumina-rich portion, and so has no significance for the application of alumina refractory.

Ternary phase diagram of $Al_2O_3-Fe_2O_3-TiO_2$ shows (Figure 6.5) that there is a distinct ternary liquidus line in the TiO_2-rich zone moving from the eutectic between Al_2O_3 and TiO_2 to the eutectic between TiO_2 and Fe_2O_3 system with decreasing temperature and has no connectivity with any other liquidus line of the system, and has no ternary invariant point or compound formation. Also, the ternary phase diagram shows a complete solid solution between two binary compounds Fe_2TiO_5 and Al_2TiO_5. The phase diagram also shows ternary liquidus formation from all the three binary systems, which are connected to form a ternary solid solution close to the Fe_2O_3 end (marked as FAss). This solid solution is encircled by ternary liquidus and has four invariant points. The FAss is surrounded by three peritectics, marked as a, b, and c in the phase diagram with temperatures 1530°C, 1520°C, and 1460°C, respectively, and one eutectic, marked as d, with temperature 1455°C. Though the ternary invariant points are low melting and liquid phase may present even up to a temperature of 1455°C in the system, all these points are close to Fe_2O_3 end and are far away from the primary phase region of Al_2O_3, and so have no significance on the alumina refractories. But, there are possibilities that low-melting phase/zone may present close to the impurity particles within the refractory and may affect properties of the impurity-rich portion.

6.6   MAIN APPLICATION AREAS

Alumina is a commonly available material and has excellent chemical, thermal and mechanical properties, both at ambient and elevated temperatures. Due to these excellent refractory qualities, alumina refractory is very common. For most of the application areas of refractories, if the specific required refractory is not available, one can use high-alumina refractory and can run the high-temperature process. But the presence of impurities in the refractory and the environment, as described, restricts its use. Table 6.7 details about the different application areas of different subclasses of alumina refractories. Obviously low-alumina-containing refractories are marginally better than fireclay refractories and used in low-temperature applications. But,

TABLE 6.7
Applications of Different Alumina-Containing Refractories

Refractory Class	Application Area
50% Al_2O_3	Preheater of cement rotary kiln; Anode baking furnace; glass tank regenerators. They are mainly used as an upgraded version of fireclay refractories. They have relatively low porosities and expansive nature up to about 1600°C, minimizing the joint gap between the bricks, and resulting in a compact lining. These refractories have low thermal expansion and good spalling resistances, and are the preferred choice for backup lining for many high-temperature industries. As the major second phase is silica, with less harmful impurities, these refractories are an excellent choice for preheaters and calcining zones of the cement rotary kilns.
60% Al_2O_3	Blast furnace stove checkers; blast furnace lining, preheating zones of cement rotary kilns. Refractories produced from calcined bauxitic clay and high-purity kaolin have a lower level of harmful impurities, resulting in low porosity, high hot strength, and creep resistance with excellent volume stability at high temperatures. Use of bauxite imparts iron oxide as an impurity and deteriorates the high-temperature properties.
62% Al_2O_3	Blast furnace hearth and tuyere; blast furnace stove checker. Low impurity and high firing results in excellent load-bearing capacity, hot strength, and creep resistance, primarily required for blast furnace hearth applications. Impurities, like iron oxide, must be minimum for hearth applications.
70% Al_2O_3	Electric arc furnace roof; steel ladle, rotary kiln. This is the most commonly used high-alumina refractory and one of the best refractories for performance to cost ratio. They are prepared mainly from calcined bauxite and high-grade clay. In many cases, the refractories are fired at a lower temperature to have lower expansions, which occur during application at high temperatures due to the reaction between alumina sources with silica available in the composition forming low, dense mullite phase associated with expansion. This secondary expansion during application is beneficial to reduce the joint size and gap between the bricks resulting in a tight and compact refractory lining, highly useful for the rotary kiln, steel ladle applications.
70% Al_2O_3 (mullite) (low iron)	Blast furnace hearth and tap hole; glass tank furnace. Special attention is given to the formation of mullite in the composition. High temperature and prolonged firing is used with reduced impurity level to complete the mullitization reaction and maximize the mullite crystal development. Excellent resistances against silicate liquids with high hot strength and creep resistances are the most important character of mullite-based refractory, which is essentially utilized in the blast furnace and glass tank furnace applications.
80%–85% Al_2O_3	Electric arc furnace roof, steel ladle, aluminum melting, and holding furnace; torpedo ladle. These refractories are mostly based on bauxite only and, in some cases, contain fine, high, pure clay or aluminas. They are also relatively common and widely used in various industries. They possess good strength and thermal shock resistances, and work well against various slag conditions. These refractories are extensively used in steel and aluminum industries.

(continued)

TABLE 6.7 (*Continued*)
Applications of Different Alumina-Containing Refractories

Refractory Class	Application Area
90% Al_2O_3	Reheating furnace hearth; carbon black reactor
	These refractories are mostly based on synthetic alumina, namely tabular and fused alumina grains, and technical alumina fines with high pure bauxite fines. They mostly consist of corundum and mullite phases, and possess high hot strength and excellent resistances against creep and chemical attack.
95%–99% Al_2O_3	Chemical, petrochemical and fertilizer industries, ceramic kilns.
	Very high, pure-alumina refractories are made from synthetic raw materials and possess only corundum as the constituting phase. Extreme-high-purity may require a very high temperature of firing, which sometimes is not practically feasible, resulting in relatively lower densification and strength properties, as that is expected. The applications of these refractories are in corrosive atmosphere of different chemical industries, mainly due to their excellent corrosion resistances coming from their purity. Use in high-temperature ceramic kilns is also important for their high temperature withstanding capacity.

higher amount of alumina containing refractories are having excellent high tempera-ture properties and are suitable for high temperature applications. Again, refractories containing >97% alumina are difficult to sinter in bulk shape under industrial and commercial conditions, and so fired products are having relatively poor qualities with respect to their alumina content. Hence, these refractories are used mostly as a pure refractory material with high chemical resistances, not for high-temperature properties. Some of the application areas of different types of alumina refractories are Figure 6.6.

SUMMARY OF THE CHAPTER

Alumina is the only compound formed between aluminum and oxygen, the third and the first most abundant elements in the earth's crust. Hence, it is abundantly available in nature but not available as pure oxide form.

The main raw material sources for alumina refractories are fire clay and all other high-alumina clays, like diasporic clay, bauxitic clay, etc.; natural aluminosilicates, namely, sillimanite, kyanite, and andalusite; bauxite and synthetic aluminas like fused alumina, tabular alumina, calcined and sintered alumina and synthetic mullite.

These raw materials, mostly in combination with one another, are used to make alumina refractories of different alumina content.

The conventional refractory-manufacturing technique of mixing, pressing, and firing is used to make the refractory shapes. Firing temperature varies with alumina content, and it is about 1750°C for very high pure refractories.

All the properties of the alumina refractories vary with the purity. The main sec-ondary phase present is silica that forms mullite in the fired product and improves the

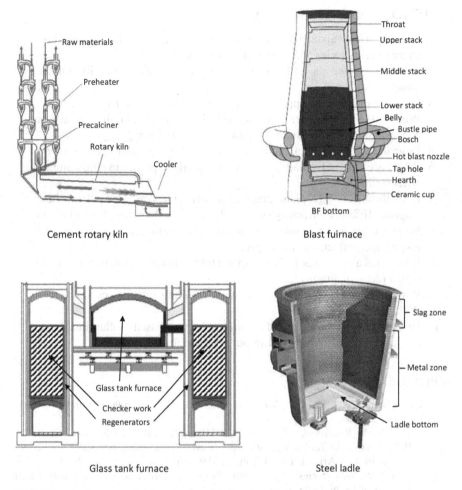

FIGURE 6.6   Some application areas and industries of alumina refractories.

refractory qualities. But other impurities, like iron oxide, titania, lime, etc. in combination drastically deteriorate the properties. Formation of low-melting compounds
in the presence of these impurities affects all the high-temperature properties badly.

There are also effects of raw material sources and their combinations on the properties of the same class (alumina content) of alumina refractories, and accordingly
their applications may vary.

Alumina refractories are applied in vivid areas with wide variation in temperatures. Low-alumina-containing refractories are a better version of fireclay refractories and are used in low-temperature applications, whereas high-alumina-containing
compositions are used in high-temperature critical application environments and
also for chemical inertness.

QUESTIONS AND ASSIGNMENTS

1. Discuss briefly about the different raw materials used for making high-alumina refractories and their common impurities.
2. Write short note on (i) bauxite, (ii) aluminosilicates, and (iii) synthetic alumina.
3. Describe in detail the applications of different high-alumina refractories.
4. What will be the variation in raw materials to make alumina refractories containing 50%, 80%, and 90% Al_2O_3 content and why?
5. Describe the complete manufacturing method of high-alumina refractory.
6. What are the advantages of having mullite phase in high-alumina refractories?
7. Draw the Al_2O_3–SiO_2 phase diagram, describe the same, and indicate in the diagram different refractory systems that are based on this phase diagram.
8. What are the major impurity phases present in alumina refractories? And discuss their effects on the properties.
9. Why silica alone is not highly detrimental in alumina refractories? Discuss with a phase diagram.
10. Why, in some cases, higher alumina-containing refractories show lower RUL values?
11. Discuss why the same class of alumina refractory based on aluminosilicates are good for cement rotary kilns but not good for steel ladle.

BIBLIOGRAPHY

1. J. H. Chesters, *Refractories- Production and Properties*, Woodhead Publishing Ltd., Cambridge, 2006.
2. C. A. Schacht, *Refractories Handbook*, CRC Press, Boca Raton, US, 2004.
3. P. P. Budnikov, *The Technology of Ceramics and Refractories*, Translated by Scripta Technica, Edward Arnold, The MIT Press, Massachusetts, US, 4th Ed., 2003.
4. A. R. Chesti, *Refractories: Manufacture, Properties, and Applications*, Prentice-Hall of India, New Delhi, India, 1986.
5. *Refractories Handbook*, The Technical Association of Refractories, Tokyo, 1998.
6. *Handbook of Refractory Practice*, Harbison Walker Refractories Company, Moon Township, PA, 2005.
7. A. O. Surendranathan, *An Introduction to Ceramics and Refractories*, CRC Press, New York, 2014.
8. *Bauxite, Indian Minerals Yearbook 2013 (Part- III: Mineral Reviews)*, Indian Bureau of Mines, Ministry of Mines, Govt. of India, New Delhi, 52nd Ed., 2015.
9. Palmer C. Sweet, Guy B. Dixon, and John R. Snoddy, Kyanite, Andalusite, Sillimanite, and Mullite, in *Industrial Minerals & Rocks: Commodities, Markets, and Uses*, Edited by Jessica Elzea Kogel, Nikhil C. Trivedi, James M. Barker and Stanley T. Krukowsk, Society for Mining, Metallurgy and Exploration Inc., Colorado, US, pp 553–560, (2006).
10. Vincent G. Hill and Errol D. Sehnke, Bauxite, in *Industrial Minerals & Rocks: Commodities, Markets, and Uses*, Edited by Jessica Elzea Kogel, Nikhil C. Trivedi, James M. Barker and Stanley T. Krukowsk, Society for Mining, Metallurgy and Exploration Inc., Colorado, US, pp 227–261, (2006).
11. I. A. Aksay and J. A. Pask, Stable and metastable equilibria in the system SiO_2-Al_2O_3, *Journal of the American Ceramic Society*, 58 [11–12] 507–512 (1975).

12. I. A. Novokhatskii, B. F. Belov, A. V. Gorokh, and A. A. Savinskaya, Zh. Fiz. Khim., *The Phase Diagram for the System Ferrous Oxide-Alumina*, 39 [11] 1498–1499 (1965).

13. T. I. Barry, A. T. Dinsdale, J. A. Gisby, B. Hallstedt, M. Hillert, S. Jonsson, B. Sundman, and J. R. Taylor, The compound energy model for ionic solutions with applications to solid oxides, *Journal of Phase Equilibria*, 13 [5] 459–475 (1992).

14. M. Kirschen, C. DeCapitani, F. Millot, J. C. Rifflet, and J. P. Coutures, Immiscible silicate liquids in the system SiO_2-TiO_2-Al_2O_3, *European Journal of Mineralogy*, 11 [3] 427–440 (1999).

15. A. I. Zaitsev, N. V. Korolev, and B. M. Mogutnov, Phase equilibria in the CaF_2-Al_2O_3-CaO system, *Journal of Materials Science*, 26 [6] 1588–1600 (1991).

16. E. F. Osborn and A. Muan, *System FeO-Al_2O_3-SiO_2, Plate 9 in Phase Equilibrium Diagrams of Oxide Systems*, The American Ceramic Society and the Edward Orton, Jr., Ceramic Foundation, Ohio, US, 1960.

17. M. Kirschen, C. DeCapitani, F. Millot, J. C. Rifflet, and J. P. Coutures, Immiscible silicate liquids in the system SiO_2-TiO_2-Al_2O_3, *European Journal of Mineralogy*, 11 [3] 427–440 (1999).

18. A. Caballero and S De Aza, Sistema Al_2O_3-TiO_2-Fe_2O_3 en aire, (The Al_2O_3-TiO_2-Fe_2O_3 system in air), *Boletín de la Sociedad Española de Cerámica y Vidrio*, 25 [2] 105–109 (1986).

19. K. Dana, S. Sinhamahapatra, H. S. Tripathi and A. Ghosh, Refractories of Alumina-Silica System, *Transactions of the Indian Ceramic Society*, 73 [1] 1–13 (2014).

7 Fireclay Refractories

7.1 INTRODUCTION

Fireclay refractories (also commonly known as firebrick) are traditionally the most common refractories and important from the total volume of production point of view. They are based on clay (hydrated alumino silicate) and contains silica and alumina as the main constituents, and so often they are also called as aluminosilicate refractories. In general, fireclay refractories contain about 25–45 wt% of alumina, and rest major component is silica. These refractories are one of the oldest ones that humanity had started using to protect heat and run a high-temperature process. But, with the advancement in knowledge, newer refractories have come up with appropriate and specific properties suitable for certain specific applications, and slowly the use of fireclay refractories is discarded. But it is one of the most common refractories that is used for any low-temperature applications and as the backup lining for many high-temperature processes. In comparison to other refractories, fireclay refractories are inferior to silica, and basic refractories in resistance to chemical attack against acidic and basic environments, respectively, and weaker both at ambient and elevated temperatures to alumina refractories.

Fireclay is a hydrated alumino silicate mineral, and the refractory is based on the alumina–silica phase diagram, as detailed in the previous chapter, Figure 6.1. The portion of the phase diagram having alumina content between 25 and 45 wt% represents the portion of fireclay refractories. As per the phase diagram the lowest temperature of liquid formation for pure fireclay, the composition is 1587°C, the eutectic temperature. However, presence of impurities strongly affects the high temperature properties of fireclay refractories, and the lowering of liquid formation temperature depends on the type and amount of the impurities present. Effect of different impurity oxides individually on silica and alumina (the components of fireclay) are discussed in Sections 5.6 and 6.5, respectively.

7.2 RAW MATERIALS AND SOURCES

Fireclay is a relatively impure secondary clay commonly found from areas close to coal mines. Although other natural deposits are also available as potential sources in many countries. Clays found close to coal mines are often got partially fired. Volatile matters of coal may come out through the porous earthy materials above the coal seams and catch fire at atmospheric conditions. This fire also burns the clay mass on the surface and the associated region. Hence, the clay mass close to the coal reserves are partially fired, and so these clays are termed as "fireclay." There is also another opinion, clays that are resistant to fire is "fireclay." Fireclay ia being used as a refractory from the very ancient times, and even as on today, other than the direct high-temperature hot-face applications, it is the most common choice, widely used

DOI: 10.1201/9781003227854-7

for making chimney and flue liners, backup lining in most of the high-temperature operations, fire-resistant pads for safety, as seen when a hearth in front of a fireplace to reduce the risk of fire.

Being a secondary clay in origin, fireclay is transported from its formation site, and acquires impurities mostly during transportation and got finer too. During transportation it also acquires impurities and organic masses. Like other clays, fireclay is malleable in raw form due to its plasticity. It can be molded, extruded, shaped by hand, and stamped. But its plasticity varies depending on its nature of origin and constituents present. As per mineralogy, it is a form of kaolinite (Al_2O_3 $2SiO_2$ $2H_2O$) and theoretically contains about 39.5% alumina, 46.5% silica, and 14% water. In loss-free basis, theoretically, fireclay contains about 46% alumina and 54% silica. But fireclay is associated with impurities like Fe_2O_3, TiO_2, CaO, etc. Due to partial firing before mining (mainly the sources close to coal mines), the water content is lower in the clay structure and varies from source to source. Hence, its plasticity is lower compared to another kaolinite type of clays. Due to this lower and nonuniform plasticity, these clays are not suitable for conventional ceramic products' manufacturing where plasticity of clay is of the prime importance. Also, the presence of impurities does not result in a good white color after firing, and so fireclay is not accepted in "whiteware industries." But for refractories, plasticity and color are of less significance. Also, the presence of lesser extent of structural water in fireclay is beneficial, as less amount of heat (and fuel) is required to remove the structural water (during calcination) for making the refractory.

When the fireclay is heated up to 500°C–600°C, kaolin minerals loose their crystallization water and an intermediate phase known as metakaolin is formed. However, this phase still exhibits a low crystalline order. The kaolin lattice does not disintegrate completely until about 950°C, and on further heating around 1000°C, mullite begins to form. Above 1100°C, only mullite, cristobalite, and glassy phases are present. The glassy phase is mainly a silica-based composition (~80% silica and only about 10% alumina, and ~5% of alkalis and alkaline earth minerals). Formation and amount of mullite phase developed in the fired products are dependent on the total alumina content of the clay. Hence, the fireclay refractory has some glassy phase in its microstructure with mullite and cristobalite as crystalline phases.

7.3 GROG AND ITS IMPORTANCE

As fireclay is a secondary clay and transported from its origin of formation by water, air, etc., it gets fine due to abrasion, erosion during transportation, and consists of very fine particles. Hence, use of only raw clay for making the refractory shape will result in huge shrinkage during drying and firing with associated dangers of cracking, warping, and breakage. So only raw clay is not used for making the refractory, and a material with coarser size fractions and similar composition but having minimum shrinkage is required to add.

A part of the clay is calcined before use, and the calcined clay is used along with the raw clay to make the final refractory composition. This prefired calcined clay is termed as "grog". Grog acts as an antishrinkage material in the fireclay composition. For better packing and compaction, we need a green mixture of refractory comprising of different fractions of particles, namely, coarse, medium, and fine particles. This grog is crushed and ground to get the desired fractions of the refractory body. Raw clay is used as the finer fraction in the composition. To make the fireclay refractory economic, crushing of already-fired broken bricks or scraps (but not contaminated) are also used as grog. Sharp angular grains of grog results in better interlocking in the shaped compositions and results in higher compaction, density, strength, and other associated properties. Grog is a major component in fireclay refractories, but the quantity of grog varies from product to product and so also the final properties. Grog content varies in the range of 20%–90%, and the rest amount is raw clay. A high percentage of grog requires highly plastic raw clay to develop proper plasticity in the mix and strength at unfired conditions. Lower grog-containing refractories have relatively lower density and strength values. Different advantages of using grogs are as follows,

1. Reduction in shrinkage (chances of warpage and cracking) of the brick on firing,
2. Higher density and lower porosity due to better packing,
3. Increase in strength values,
4. Less requirement of water for mixing.

7.4 BRIEF OF MANUFACTURING TECHNIQUES

Mined raw clay is first moistened and kept in an open atmosphere in thin layers for decaying of organic matters present in it. This helps to form some organic acids (humic acid) from the organic masses present in it and increases the plasticity. This process is called "weathering or souring." Weathered clays are then crushed and ground. Mainly the raw clay is used as a fine fraction to utilize its plasticity in the refractory composition. This clay is mixed with the grog fractions, as per the total grog content. A blend of two or more clays is normally done for the manufacture of fireclay refractories to accommodate any variation in any of the clays without much variations in the processing. No separate green binder is required, and used as raw clay itself is a plasticizing and bonding material in the green stage.

The mixture of different grog fractions (coarse, medium and fine) and fine raw clay is mixed either in dry or wet conditions. In dry conditions, the ingredients are directly charged in the mixer (generally pan mixer) and mixed with the minimum amount of water required for pressing. In wet mixing, first the mix with the raw clay is soaked with water for about 2 days, so that fine clay can be distributed uniformly all through the batch composition, and then the mix was charged in pan or pug mill. The mixed composition is allowed to be aged for days in a cool cellar for better aging and increased uniform plasticity.

The shaping of the mixed refractory compositions is done by hand molding, pneumatic ramming, or pressing. Depending on the shaping process, the moisture content of the mixed composition may vary between 4 and 12 wt%. Lower moisture content requires high molding pressure, and higher moisture content results in a higher chance of slumping. Also, low-pressure products are less dense and have lower strength compared to dry pressed products. Again, hand molding and ramming processes produce nonuniform pressure during shaping and results in nonuniform properties. The size of these low-pressure products may also not be appropriate due to high shrinkage, and so refractories with greater dimensions are made and sintered, which are cut to get accurate dimensions. Again pressing method allows a very high production rate, but cannot produce any complicated shape and large-sized product. Increasing demand for greater productivity results in most of the products shaped by the pressing method, employing various types of presses including hydraulic presses. Uniform and homogeneous structure, better compaction, higher density, improved strength, etc. are the characteristics of these products. A compaction pressure of 50–80 MPa is used for the fireclay refractories. Again the fireclay raw particles are very fine and result in high shrinkage both after drying and firing. Hence, the dimensions of the molds and dies are to be adjusted accordingly.

Shaped articles are dried before firing to remove mainly the physically absorbed moisture, and drying is critical for fireclay bodies. Very slow drying is done with no direct heating of the surface in the initial periods. Very fine clay particles may shrink unevenly, and may result in nonuniform drying and nonuniform shrinkage causing cracks. This is more important for high moisture-containing shapes. Generally, 1– 4 days of natural drying in open air (not in direct sunlight) is done before a final drying in the ovens or driers. Hot air or waste heat is used in the driers to reach the highest temperature of 150°C–250°C, and a total drying time of 24–200 hours are used depending on the moisture content and grog amount.

After drying, the dried shapes are fired in batch (down draft, chamber kiln) or continuous kilns (tunnel kiln). Firing is done slowly to accommodate the structural changes that occur within fireclay, reactions, and phase formations and shrinkage. The total firing schedule is classified into five stages, depending on the changes occurring the refractory being fired and the temperature zone.

In the first stage, from room temperature to about 500°C, most of the moisture is removed from the shapes. Any leftover of physically absorbed water after drying is expelled around 150°C–200°C, and the chemically absorbed water is removed around 400°C–450°C. This stage is called "steaming" or "smoking," as steam or smoke is observed to be coming out from the furnace. Steaming is continued for about 20–24 hours. All the moisture present as physical or chemical water are removed, and the raw clay undergoes a nonreversible structural change. It becomes an anhydrous aluminosilicate, which is termed as metakaolin. Next, the second stage is termed as "decomposition," associated with various decompositions and reactions that occur within the fireclay, mainly due to the presence of the impurities. This stage continues up to about 900°C for about 20–24 hours. Decompositions of various carbonates (calcium, magnesium), sulfates (iron, calcium), sulfides (iron) occur during this temperature range, and any free organic mass is finally removed, any free carbon or sulfur will be oxidized, etc., will be completed. The third stage is named as "full firing," where no such decomposition is occurring, and not much structural change occurs within the refractory. The only major change that occurs in this stage is the decomposition of metakaolin to form mullite and excess silica. This reaction is an exothermic one and results in the evolution of heat. This stage is considered up to about 1300°C and lasts for up to 36 hours. Above 1300°C, liquid phase starts forming due to the presence of various impurities and their reactions with the silicates. This fourth stage is known as the "incipient fusion." In this stage, initially mullite starts crystallizing, and clay particles get hardened. Then the liquid phase helps in enhancing the sintering and fill up the gaps between the hard grog particles. The pores are filled with the liquid phase, shrinkage occurs, and a compact, hard, dense mass is produced. This stage may continue up to 1500°C, and this is the last stage of heating. Next the refractories are cooled, and the stage is called "cooling or annealing." Slow cooling is done to crystallize the liquid phase and enhance the strength of the refractory. The highest temperature of firing is generally above the usual application temperature of the refractories to avoid any further shrinkage (PLCR) during use. Also, a higher rate of heating or cooling may be adapted for the refractories having high grog content. Fewer grog compositions are having more amount of finer raw clay, which results in higher shrinkage.

Lower heating rates are used for refractories containing lesser grog, as they undergo higher shrinkage. After firing, the fireclay refractories consist of mullite, cristobalite, residual quartz, and glass. In fired bricks, the mineral components (phases) are not present at the equilibrium conditions and their amount changes with further heating. Only after the installation and use of the fireclay refractories in the furnace, phases present, and their amounts may change due to high temperature and approach the equilibrium conditions. At higher temperatures and longer holding periods, the mullite content changes little, whereas the content of cristobalite and quartz decreases and finally disappears totally at 1400°C–1500°C. The fireclay bricks then consist of clay, mullite, and a viscous glassy phase. This glassy phase consists silica, alumina, alkalis, and other fluxing agents coming from the impurities present in the composition. Over-fired refractories may deform due to the presence of excess liquid phase. Under-fired refractories will have lower strength and can be fired again for the attainment of the properties. Fast firing may result in "black heart", a defect due to insufficient availability of oxygen for combustion and incomplete decompositions of the impurities inside the refractory during firing, causing rejection of the refractory.

7.5 CLASSIFICATIONS AND PROPERTIES

The fireclay, being a natural material and a clay of secondary origin, varies widely in composition and properties. These variations are prominent when the sources of the raw material are varying. The Al_2O_3 content in naturally occurring fireclay varies between 25% and 45%, and the rest being silica as the major component with impurities. Hence, the refractory made up from fireclay is also having varying properties depending on the alumina and impurities content. The usefulness of fireclay refractories is largely due to the presence of mineral mullite (containing 72% Al_2O_3), which forms during firing and having properties, like high refractoriness, excellent resistance against silicate environment, thermal shock resistance, and low thermal expansion. In general, higher the alumina content in fireclay better will be the high-temperature properties. Hence, fireclay refractories are mainly classified as per the alumina content, and all the different properties also vary with the classification. The conventional nomenclature is referred to the "heat duty" of the refractory, which indicates the temperature withstanding capability of the material in qualitative

TABLE 7.1

Properties of Fireclay Refractories as per Their Heat Duty (Al$_2$O$_3$ Content)

Properties	Heat Duty			
	Super	High	Moderate	Low
Al$_2$O$_3$%, minimum	40	38	30	25
Fe$_2$O$_3$%	1.0–1.5	1.5–2.0	2.0–2.5	2.5–3.0
Size tolerance	1	2	2	2.5
Apparent porosity (%), maximum	20	25	26	28
Cold crushing strength (MPa), minimum	20	25	27	22
PCE (ASTM) No, minimum	33	32	30	23
RUL, Ta (°C), minimum	1450	1400	1370	1340
PLCR (%)	0.4 at 1450°C	1.5 at 1450°C	1.0 at 1350°C	

terms. Higher alumina content allows the material to withstand higher temperature (increases fusion/liquidus temperature), thus improves all the high-temperature properties. Table 7.1 shows the properties of different fireclay refractories as per their classification based on heat duty.

7.5.1 SUPER HEAT DUTY (AL$_2$O$_3$ CONTENT ~ 40%–45%)

Super heat duty fireclay refractories have a minimum alumina content of 40 wt% and show good strength and refractoriness, high volume stability at high temperatures, and superior thermal shock resistances. Their refractoriness is close to 1800°C and has an RUL value of 1450°C (minimum). Higher alumina content requires a higher temperature of firing for proper densification, and the high-temperature firing enhances the high-temperature strength, stabilizes their volume and mineral composition, increases their resistance to corrosive agents, and makes them highly resistant to disintegration by carbon monoxide gas.

7.5.2 HIGH HEAT DUTY (AL$_2$O$_3$ CONTENT 35%–40%)

These refractories are having alumina content about 38% (35%–40%) with a refractoriness around 1700°C and a minimum RUL value of 1400°C. They are important for their greater resistance to thermal shock and used mainly in those linings operated at moderate temperatures over long periods of time but subject to frequent shutdowns (thermal shocks).

7.5.3 MEDIUM HEAT DUTY (AL$_2$O$_3$ CONTENT 30%–35%)

Medium duty fireclay refractories contain alumina minimum of 30 wt% and have refractoriness in the range of 1650°C with RUL about 1370°C. Due to strong bonding from the liquid phase, these refractories can withstand abrasion better.

7.5.4 Low Heat Duty (Al₂O₃ Content 25%–30%)

Low heat duty fireclay refractories have an alumina content about 25%–30%, with a refractoriness of about 1600°C and RUL of 1340°C. They are mainly important as backup bricks to support many high-temperature applications.

7.5.5 Semi-Silica

Semi-silica fireclay refractories contain 18%–25% alumina and 65%–80% silica, with a low content of alkalis and other impurities. Though these refractories have very high silica content, they are still considered as a type of fireclay refractories. They have a refractoriness of 1650°C and RUL of ~1400°C. These refractories have less shrinkage and PLCR values, and also have excellent load-bearing strength and volume stability within its application temperatures.

The properties of fireclay refractories depend on the amount and type of clay and grog used, amount of alumina and other impurities present in the composition, types of processing involved, etc. Impurities like alkalis and alkaline earth oxides, iron oxide, titania, etc. react with the silicate phases and reduce the liquidus temperature of the compositions; thus, they affect the high-temperature properties like refractoriness, RUL, shrinkage, etc. The softening behavior of the fireclay refractory is dependent on the amount and the composition of the glassy phase. Moisture content used for shaping the refractories affect the properties by the creation of pores. The amount of raw clay also affects the shrinkage and amount of moisture required for processing, and so it also affects nearly all the properties of the fired refractories.

As silica is present as a major component, fireclay refractories behave little acidic in chemical nature. Hence, they are weak against any basic slag, fume, fluxes, and environments. Compact and less porous microstructure of the refractories are stronger against corrosive attacks. Again mullite being a major phase in the fired refractory, which has a very low thermal expansion characteristics, fireclay refractories also show relative low thermal expansion properties and, in turn, produce better thermal shock resistance character. Coarse textured and high-grog-containing compositions again have better thermal shock resistance properties than those of fine textured and low grog-containing ones.

Resistance against carbon monoxide gas is specially very important for fireclay refractories. The details have been discussed in Chapters 3 and 4. Fireclay refractories must have low iron content for the application in the stack area of blast furnaces, so that they are strong against such degradation.

7.6 MAIN APPLICATION AREAS

History of refractories indicates that initiation of these materials started with fireclay-based refractories. With the advancement of human civilization, the use of fireclay-based refractories have increased tremendously, and until the early of the 19th century, it was the most used refractory material. But with the advancement of science and technology and specific demand for refractory properties for specific application areas, slowly and gradually the use of fireclay refractory got reduced. Also, fireclay refractories are acidic in nature, which restricts its wide applicability in other environments. But, still being a comparatively cheap refractory, it is the most common preferred item for any low-temperature applications.

Other than the application environment, the use of fireclay refractories is also influenced by several other parameters, like the temperature of applications, load and strength factor, volume stability, thermal shock, etc. The major application areas of fireclay refractories are as follows.

Upper stack of blast furnace: This application does not demand a very high-quality refractory, as temperature varies between 300°C and 800°C (as shown in Figure 6.3), so common and economic fireclay refractories suit here. But the critical part is its carbon monoxide environment. Fireclay refractories are suitable for such applications with a maximum Fe_2O_3 content of 1.5%. Depending on the temperature of application area, the alumina content may vary between 35% and 45%.

Hot blast stove: Refractories used in this application require high creep resistance, volume stability, and thermal shock resistance (as shown in Figure 5.5). Low-creep fireclay refractories with alumina content around 40%–45% and iron oxide below 2% are used.

Cement rotary kiln: Fireclay refractories are suitable for preheater cyclones, pre-heating zones, calcination zone, etc., for cement rotary kilns (as shown in Figure 6.3). Temperature varies in these zones from about 300°C to 1100°C, and depending on the temperature alumina content varies between 25% and 45%, and iron oxide is restricted below 2.5%.

Glass tank furnace: Fireclay refractories are used in the different applications of glass industries. Due to high creep resistance up to about 1300°C and excellent thermal shock resistance, they are the preferred choice for checker work of the regenerators of glass tank furnace (except top part), 40%–45% alumina-containing fireclay with iron oxide < 1.2% are useful for regenerators wall, rider arch applications (as shown in Figure 6.3). Similar alumina content with iron oxide ~2% is useful for regenerator wall, lower bottom layer, and backup of melting tank, etc.

Baking furnace of aluminum industries: These furnaces are large in structure, built underground with a floating foundation, and are made up of several pits or cells. High-grog fireclay refractories with alumina content 40%–45% and iron oxide ~2% are used in the flue duct. Depending on the temperature and service conditions, backup walls and head walls are lined with fireclay refractories with alumina 35%–40% and iron oxide up to 2.5%.

Metal casting: Fireclay refractories are also used in steel-/metal-casting industries, mainly as a funnel, runner, cast pipe, center brick, end runner brick, ingot mold bottom plate, etc. 30%–40% Al_2O_3-containing refractories are used for these applications.

Other than the above applications, fireclay refractories are also used in heat-recovery systems of different industries (regenerators and recuperators), annealing and reheating furnaces of steel industries, carbon roasting furnaces, sugar industries, and all the typical and low-temperature applications like fireplaces, ovens, flue lines, chimney linings, etc. (Figure 7.1).

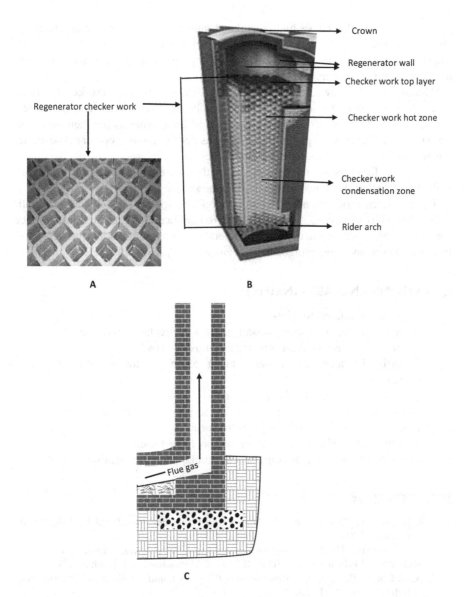

FIGURE 7.1 Applications of fireclay refractories, (a) specific stacking of fireclay refractories in regenerator, (b) regenerator with the stacked checker work, and (c) chimney.

SUMMARY OF THE CHAPTER

Fireclay refractory is the oldest refractory and one of the most widely practiced refractory material.

Natural fireclay is the main raw material for this refractory. Fireclay is a secondary clay commonly found close to coal mines and often with less crystalline water due to partial natural firing from the coal volatile matters.

Very fine particles cause high shrinkage and cracking of the fired refractory. Hence a prefired clay mass, grog, is used as antishrinkage material to avoid these. The addition of grog also provides advantages like enhancement of density, strength, and reduction in water demand during mixing.

The firing of fireclay refractories is little critical due to the presence of various moisture, gas evolution, and structural changes. It also takes a long firing schedule.

Commonly, fireclay refractories are of four different types as per their heat duty (maximum application temperature) classified as per the alumina content. The higher the amount of alumina, the better the hot properties.

Generally, fireclays are medium dense, medium strength, little acidic refractory with good thermal shock resistance and volume stability.

Fireclay refractories are most commonly preferred for any low-temperature application due to its cheapness. The most preferred application areas are blast furnace stack, hot blast stove, regenerator and recuperators, aluminum baking furnaces, preheater cyclones, and preheating zones of cement rotary kilns, etc.

QUESTIONS AND ASSIGNMENTS

1. Discuss the mineral fireclay?
2. What are the advantages and disadvantages of fireclay as refractory?
3. What is grog, and what are advantages of using grog?
4. Describe in detail the manufacturing technique for making fireclay refractories.
5. Discuss the firing of fireclay refractories.
6. Detail the classification of fireclay refractories with the property variation.
7. Discuss the properties of fireclay refractories.
8. Briefly mention the applications of fireclay refractories.
9. Describe the variation in properties of fireclay with the alumina content.

BIBLIOGRAPHY

1. J. H. Chesters, *Refractories- Production and Properties*, Woodhead Publishing Ltd., Cambridge, 2006.
2. P. P. Budnikov, *The Technology of Ceramics and Refractories*, Translated by Scripta Technica and Edward Arnold, The MIT Press, Massachusetts, US, 4th Ed., 2003.
3. A. R. Chesti, *Refractories: Manufacture, Properties, and Applications*, Prentice-Hall of India, New Delhi, India, 1986.
4. *Refractories Handbook*, The Technical Association of Refractories, Tokyo, 1998.
5. *Handbook of Refractory Practice*, Harbison Walker Refractories Company, New York, 2005.
6. D.N. Nandi, *Handbook on Refractories*, Tata McGraw-Hill, Noida, India, 1987.

8 Magnesia Refractories

8.1 INTRODUCTION

The major part of the total refractory produced, more than two-third fraction, is used in iron and steel industries. Again, the environment in steel industries is basic since long, and to make the refractories chemically compatible with the steel-making process, basic refractories are must for metal-contact areas in steel manufacturing plants. Alumina, being neutral in character, survives well in the iron and steel industries, but as it shows little acidic character at high temperatures and reacts with lime present in the slag, alumina is not the most suitable refractory for such applications. Hence, use of basic refractories is of prime importance, especially in those liquid contact areas having high basicity.

As per basicity point of view lime, CaO (melting point 2570°C) is the best among the oxides, and the lime refractory may be the most suitable one for any high-temperature and high basic applications. Also, lime is abundantly available in nature, calcium being the fifth most abundant element in earth's crust. But, among all these positive qualities, only the problem of hydration has restricted lime to be commercially successful as refractories. Lime has very high hydration character, and it reacts with the moisture of air instantaneously forming calcium hydroxide $[CaO + H_2O = Ca(OH)_2]$ even in sintered or a fused form. This hydration causes expansion and crack in the sintered shape, and results in formation of powdered hydroxide product. Plenty of research work has been done to restrict or reduce the hydration character of lime, but the commercial and economical process to do the same is yet to be established, and lime refractories are not being practiced commercially and widely.

As per basicity and availability point of view, next to lime comes the magnesia. Magnesia or magnesium oxide (MgO) is a white hygroscopic solid mineral that is available in the natural state in limited amount with a mineralogical name "periclase." It also absorbs moisture from the environment and forms magnesium hydroxide $[MgO + H_2O = Mg(OH)_2]$, but this reaction is relatively sluggish compared to lime, and high sintered and fused products have very limited hydration tendencies.

In the earth's crust, magnesium (Mg) is the eighth most abundant element and constitutes about 2% of the earth's crust. Although magnesium is available in more than 60 minerals, only dolomite, magnesite, brucite, carnallite, and olivine are of commercial importance due to its concentration level. Again, magnesium is the third most available element dissolved in seawater, with a concentration averaging 0.13%. Magnesium and magnesium-bearing compounds can be produced from seawater and brines of wells and lakes.

Relatively high basic character with wide natural availability and limited hydration character have made magnesia the most suitable material for basic refractories. In most of the application areas with high basic environments, especially the slag

DOI: 10.1201/9781003227854-8

contact areas of steel-making and steel-processing applications, refractories containing magnesia is most suitable and being practiced globally.

8.2　RAW MATERIALS AND SOURCES

Magnesia refractories are manufactured from various raw material sources that are described below.

8.2.1　MAGNESITE

Magnesite ($MgCO_3$) is the naturally occurring carbonate of magnesium and the most common source of magnesia used for many important industrial applications. The term "magnesite" literally refers only to the naturally occurring mineral; however, in common usage, it is being used for the end products as well (like magnesite refractories that actually contain MgO). Theoretically, magnesite contains 47.6% MgO and rest carbonates. It is available in two different physical forms in nature, namely, (i) macrocrystalline and (ii) cryptocrystalline. Cryptocrystalline magnesite is generally of a higher purity than macrocrystalline ore but tends to occur in smaller deposits than the macrocrystalline form. It is commonly associated with calcium carbonate, ferrous carbonate, and magnesium silicates (like serpentine and olivine).

At high temperatures, $MgCO_3$ decomposes to magnesium oxide and carbon dioxide.

$$MgCO_3 \rightarrow MgO + CO_2$$

The decomposition starts at a temperature of around 350°C, but heating to around 900°C is required to obtain complete oxide phase due to interfering readsorption of liberated carbon dioxide. This low-temperature calcined MgO can be hydrated from the atmospheric moisture and form magnesium hydroxide. To reduce or avoid hydration tendency, calcination is done between 1100°C and 1300°C to get caustic magnesia obtained in a loose form. Caustic calcined MgO is characterized by high chemical reactivity, fine crystal size, high porosity, and low bulk density. Caustic calcined MgO occasionally finds use in the metallurgical industry as slag conditioner, but rarely employed in refractory products and never as a major aggregate ingredient. It is used in agricultural and industrial applications, like feed supplement to cattle, fertilizers, electrical insulations, industrial fillers, and in flue gas desulfurization.

This calcined magnesia is fired to very high temperature, to make it completely nonreactive, and is called dead burnt magnesia (DBM). Before dead burning, magnesite is beneficiated by conventional methods (optical sorting, heavy media, and magnetic separation) or by processes like flotation to reduce the impurity content. The calcined form of magnesia is briquetted and then dead burnt in a rotary kiln or shaft kiln at temperatures around 1900°C–2000°C. MgO content in DBM varies between 85% and 98%, and the major impurities are SiO_2, CaO, Fe_2O_3, etc. DBM is characterized by low chemical reactivity, large crystal size, low porosity, and high bulk density. Magnesia is the highest-melting common refractory oxides and is the most suitable heat-containment material for high-temperature processes in the steel industry. And DBM is the most widely and commonly used source of magnesia. Although China is the largest producer of DBM, there are significant resources in North Korea, Turkey, Austria, Slovakia, Brazil, Russia, Spain, and Australia.

8.2.2 FUSED MAGNESIA

Fused magnesia is produced in an electric arc furnace using high-grade magnesite or calcined magnesia as raw materials. About 12 hours is required for the fusion process at temperatures above the melting point, 2825°C, and then it is cooled very slowly. The process promotes the growth of very large crystals of periclase (even greater than 1000 μm compared to less than 100 μm for dead burned/sintered magnesia), with a density approaching the theoretical density of 3.58 g/cc. The resulting product is then crushed and ground to get the desired fractions to be used for refractory manufacturing. Generally, fused magnesia is superior to dead burnt magnesia in strength, abrasion resistance, and chemical stability. Fused magnesite is manufactured in the locations where natural magnesite is in abundance, e.g., China, Turkey, and Australia, etc.

The addition of fused magnesia grains can significantly enhance the performance and durability of the basic refractories. The primary reasons are higher density, large periclase crystal size, and realignment of accessory silicates. Also, as the magnesia is fused above 2800°C, it becomes immune to react at lower temperatures, and fused mass also has lower porosity levels. So fused products have lesser tendency to react in any environment. Refractory-grade fused magnesia is normally characterized by:

- high magnesia content (minimum 97% MgO)
- low silica content (lime-to-silica ratio >2)
- densities ~3.50 g/cc or more
- large periclase crystal sizes (1000 μm minimum)

Due to its relatively high chemical stability, strength, resistance to abrasion, and excellent corrosion resistance, refractory-grade fused magnesia is used in high corrosive and wear areas in steel making, like the molten-metal-contact areas. Fused magnesia also has high thermal conductivity. But the main constraints for making fused MgO are the size and number of electric arc furnaces and the cost of energy. The manufacturing of fused magnesia is highly power intensive with electricity consumption varying between 3500 and 4500 kWh/ton.

8.2.3 SEA WATER MAGNESIA

Magnesium is present as a soluble salt in seawater, which contains about 1.3×10^{-3} kg/L Mg^{2+} ions in combination with chloride and sulfate ions. When it is represented in the form of an oxide (MgO), this amount becomes $\sim 2 \times 10^{-3}$ kg/L. Theoretically, this can be extracted easily by the addition of a suitable alkaline compounds in sea water. The first large-scale commercial plant for producing refractory-grade magnesia was built at Hartlepool, UK, by the Steetley Company in 1938, using seawater and dolomite. Because of their consistent composition and quality, seawater magnesia is preferred compared to natural magnesia. The impurities, present as trace elements in sea water, are controlled by strict adherence to processing conditions and chemical additives. By controlling the preparation and precipitation conditions, the amounts of chloride, sulfate and other ions can be controlled. Also, certain trace elements are added to this magnesia for the betterment in sintering and properties.

To precipitate out magnesium from seawater, nearly all the processes use lime-bearing compounds like calcined and hydrated limestone or dolomite. Calcium ion replaces the magnesium ion, combines with the chloride or sulfate anions, and gets dissolved as soluble salt, whereas magnesium precipitates out in its hydroxide form. In this process, first limestone or dolomite is calcined in a rotary kiln at 1300°C–1400°C.

$$CaCO_3 = CaO + CO_2$$

$$(Mg, Ca)\ CO_3 = MgO.\ CaO + CO_2$$

Then the calcined product, lime or doloma, is hydrated.

$$CaO + H_2O = Ca(OH)_2$$

$$(Mg, Ca)\ O + H_2O = (Mg, Ca)(OH)_2$$

This hydrated material is then reacted with seawater stored in big tanks with agitation, which results in the following chemical reactions.

$$MgCl_2\ (\text{from seawater}) + Ca(OH)_2 = Mg(OH)_2 + CaCl_2\ (\text{dissolves in seawater})$$

$$MgCl_2\ (\text{from seawater}) + (Ca, Mg)(OH)_2 = 2Mg(OH)_2 + CaCl_2\ (\text{dissolves in seawater})$$

$$MgSO_4\ (\text{from seawater}) + Ca(OH)_2 = Mg(OH)_2 + CaSO_4\ (\text{dissolves in seawater})$$

$$MgSO_4\ (\text{from seawater}) + (Ca, Mg)(OH)_2 = 2Mg(OH)_2 + CaSO_4\ (\text{dissolves in seawater})$$

$Mg(OH)_2$ precipitates out from the water and settles down at the bottom of the tanks with the spent sea water flowing over the top. A flocculant is used to make the settling process more efficient. Part of the slurry is sent back to the reactor to act as seeding material, and the rest goes to secondary thickeners, where the precipitate is further thickened before filtration. Finally, the slurry is filtered on rotary vacuum disk filters and additions (may be lime, silica, and iron) are made to control the ratios of the impurities and properties enhancement of magnesia. The filter cake, containing about 50% solids, is then either fed directly into the rotary kiln and fired above 1900°C, or calcined first around 1100°C, then cooled, pelletized, and sintered in the shaft or rotary kilns above 1800°C.

Very high temperature (over 1800°C) is used for sintering (dead burning) of magnesium oxide and results in the formation of larger periclase (MgO) crystals. The kilns transform low-density magnesium oxide into high-density sinters, which has a density of about 3.4 g/cm³. This product has a buff-brown color and is chemically less reactive. The sintered magnesia produced through this process can also have varying MgO contents, between ~90% and 98%; however, if a high-purity magnesia source is used in this process, the products can have MgO contents in excess of 99%. Also, for making high-purity magnesia calcined dolomite is not used, as natural dolomite contains about 1.3% Fe_2O_3, which will impart higher iron impurities in the product.

The main impurities of seawater magnesia are lime, silica, boron oxide, etc. Calcium bicarbonate present in seawater precipitates out as calcium carbonate on reaction with calcium hydroxide.

$$[Ca(HCO_3)_2 + Ca(OH)_2 = 2CaCO_3 + 2H_2O],$$

which remains as CaO impurity in the sintered product. Acid pretreatment of seawater can reduce this lime impurity. Also, any unconverted hydrated calcined lime of dolomite used may result in lime impurity. Most of the silica impurity comes from the sand suspended in seawater. Boron is present in small amounts in seawater as boric acid (seawater contains about 15 ppm, expressed as B_2O_3 content). The precipitates of magnesium hydroxides have a high tendency for absorbing boron, and the final concentration in the oxide can be as high as 0.4%. This boron contamination can be reduced by using an excess of lime but may result in excess lime impurity in the product. Also, removal of boron can be done during the dead-burning stage by volatilization. This volatilization rate can be accelerated by the addition of certain alkali metal salts, e.g., potassium hydroxide. The presence of boron in magnesia is highly detrimental for its hot properties, as it produces very-low-melting magnesium borate phases, like $Mg_3B_2O_6$, having melting point of 1360°C.

Due to the presence of salts, the specific gravity of sea water is higher and commonly varies between 1.02 and 1.03. The highest concentration of mineral salts is available in the water of Dead Sea, where the water has a specific gravity of 1.24. Due to the high concentration of minerals, many of the sea water manufacturing plants are located near the Dead Sea, and countries like Israel, Jordan is a major producer. Again, as the river water has a much lower salt concentration, the river connected to neighboring sea areas are having significantly low salt concentration and so are not a good option for making sea water magnesia.

8.2.4 MAGNESIA FROM NATURAL BRINE SOURCE

Magnesia is also produced from brine solution available in the surface (lakes), subsurface, and subterrain deposits. Composition and concentration of the brine depend on the chemistry and soluble mineral percentage of the surrounding salt-bearing rocks. Naturally occurring subsurface brines are available in porous sandstones and other porous rocks that support the composition. Generally, the concentration of magnesium ion in brine is much higher than seawater, and so the amount of water required to be processed is significantly lower than the sea water plants. Hence, the same-capacity magnesia-producing plant requires much smaller tanks and other processing equipment. The most important sources of subsurface brine are available in Nederland, US (Mississippian and Pennsylvanian beds, Michigan, etc.).

Lake and well sources of brine are easy ones, as the extraction is not so complicated, and involves only the pumping of the solution to the processing plant. However, the underground recovery is a complicated process. Leaching or solution mining of the salt available under the earth using drilling or recovery technology is used as practiced in oil and gas industries. A wide vertical bore well is first drilled from the surface to the deposit. Several concentric tubes or casing are used for the liquid flow to and from the different layers of the deposit source. Water is pumped in which dissolves the present salt underneath, and the solution comes out to the top under the pressure of incoming water.

The brine solution thus obtained is processed in a similar way as that of the sea water, and the rest process is exactly similar to that of the sea water magnesia manufacturing. However, composition and purity of the final magnesia vary from sea water products, as many impurity salts are dissolved in a brine solution, coming from the surrounding rocks of the brine source.

8.2.5 CHARACTERISTIC OF THE RAW MATERIALS AFFECTING THE REFRACTORY

The important characteristic of the magnesia raw materials that affect the properties of the final refractory is purity and impurity types and content, crystallite size, density, etc. It is the overall purity of the raw material that plays an important role in determining the MgO content of the final refractory and judges the suitability for a specific application. Again impurities, depending on the type and amount, form low-melting liquid phases during use at high temperatures. These liquids when present in-between the grains slide them and finally results in degradation of high-temperature strength. Also, the liquids help corrosive slags to invade the refractory and drastically reduce the corrosion resistance. Impurities that form a liquid phase at lower temperatures are more detrimental, and among them SiO_2 and B_2O_3 are important. Among different impurities, SiO_2, CaO, Al_2O_3, Fe_2O_3, and B_2O_3 (in sea water MgO) are important and common. The impurities do not exist separately and affect MgO individually, but they are present in combination, and their combined effect at high temperature is more detrimental to MgO.

The crystallite size of MgO is also an important parameter for the corrosion of the refractory, as it determines the available surface area for any reaction. Any chemical reaction is a surface activity, and reduction in the available surface by increasing the crystallite size of magnesia grain will enhance the corrosion resistance. Fused and high-fired magnesia are better in corrosion resistance mainly due to their increased crystallite size. The higher density of the starting materials indicates a lower total porosity value (open and closed pores), and reduces the chance of corrosive liquid penetration and corrosion. It is considered that the densest body offers the best resistance to corrosion against slags, and is also the strongest to resist abrasion. Hence high-dense raw materials will result in better properties.

8.3 BRIEF OF MANUFACTURING TECHNIQUE

Dead burnt, or sintered magnesia is used for the manufacturing of the refractories. Fired lumps are crushed and ground to obtain desired size (max ~5 mm), and different fractions are used for obtaining the highest compaction density using coarse, medium, and fine. Crushing and grinding may impart impurities like dust, iron particles, which are required to be cleaned before using the fractions in the mixer.

Different fractions are mixed with a bonding material required to provide bonding between the hard-fired magnesia particles. This bonding material may be clay, sodium silicate, milk of lime, etc., depending on the type of raw materials, impurities present, and desired properties in the fired refractories. Also, very little (~0.5%) mill scale (a mixture of different iron oxides) is used to enhance the sintering and strength. In certain applications, 5%–6% alumina is also added as fine material to form magnesium aluminate spinel within the magnesia refractory matrix, especially to improve the thermal shock resistance. The mixture is mixed with 4%–7% of water (sulfate lye is also added for green strength) to obtain proper consistency, uniformity, and suitable plasticity for subsequent shaping process. Shaping is done in a hydraulic press at a specific pressure of 80–120 MPa. Next, the shapes are dried and fired. Drying of magnesia refractories is critical, as cracks may occur during drying process due to,

1. use of soft fired magnesia, which may hydrate,
2. presence of excessive fines causing shrinkage and hydration, and
3. high temperature of the drier.

Defect- and crack-free samples are further processed for firing. Firing is done in batch-type (chamber) or continuous-type (tunnel) kilns. Firing temperature ranges from 1550°C to 1800°C, depending upon the purity and impurities present. Higher the temperature better may be the properties, as the shrinkage during application will be reduced. A soaking time of about 4–8 hours is given at the peak temperature for uniformity in temperature and uniform property development. As per molecular bonding point of view, magnesia has a strong bonding (ionic) character, compared to silica and alumina, so it has a lower diffusivity and poor sinterability. Hence, magnesia sinters poorly, and sintering aid (mill scale) is necessary. Iron oxide of mill scale reacts with magnesia and forms magnesio-ferrite, which is present in between the magnesia grains, and helps in bonding between the magnesia grains and improves sintering, densification, and strength.

8.4 EFFECT OF LIME: SILICA RATIO

Both lime (CaO) and silica (SiO_2) are common and abundant impurities for magnesia refractories. The presence of lime is not as harmful as most of the impurities because alone it does not degrade the properties of magnesia (as detailed in the phase diagram, described later). But silica, if present, produces low-melting silicates and affect the high-temperature properties drastically. Impurities are more effective in finer form, due to the greater reaction from higher surface area and produces liquid phase at low temperatures. So the deterioration in properties starts mainly in the matrix phase. But the deteriorating effect of silica can be controlled by the presence of lime, as the reaction compounds form in the presence of lime

TABLE 8.1

Effect of Lime–Silica Ratio on Compound Formation in Magnesia Refractories

Lime : Silica (CaO:SiO$_2$) Ratio, X		Possible Phases	Chemical Formula	Melting/Fusion Point (°C)
Molar ratio	Weight ratio			
X < 1	X < 0.94	Forsterite	2 MgO.SiO$_2$	1890
		Monticelite	CaO.MgO.SiO$_2$	1503
		Magnesio-ferrite	MgO.Fe$_2$O$_3$	1760
X = 1	X = 0.94	Monticelite	CaO.MgO.SiO$_2$	1503
		Magnesio-ferrite	MgO.Fe$_2$O$_3$	1760
1 < X < 1.5	0.94 < X < 1.4	Merwinite	3 CaO.MgO.2 SiO$_2$	1550
		Monticelite	CaO.MgO.SiO$_2$	1503
		Magnesio-ferrite	MgO.Fe$_2$O$_3$	1760
X = 1.5	X = 1.4	Merwinite	3 CaO.MgO.2 SiO$_2$	1550
		Magnesio-ferrite	MgO.Fe$_2$O$_3$	1760
1.5 < X < 2	1.4 < X < 1.87	Merwinite	3 CaO.MgO.2 SiO$_2$	1550
		Belite	2 CaO.SiO$_2$	2130
		Magnesio-ferrite	MgO.Fe$_2$O$_3$	1760
X = 2	X = 1.87	Belite	2 CaO.SiO$_2$	2130
		Magnesio-ferrite	MgO.Fe$_2$O$_3$	1760
X > 2	X > 1.87	Belite	2 CaO.SiO$_2$	2130
		Alite	3 CaO.SiO$_2$	2070
		Magnesio-ferrite	MgO.Fe$_2$O$_3$	1760

are different. Hence, the amount of lime present and the ratio of lime:silica is important for the properties and performance of the magnesia refractories. Other impurities, commonly iron oxide, also take part in the reactions and form complex compounds.

Again, if the amount of lime impurity is high and free lime is present (not reacted with silica or other impurities), then the free lime can deteriorate the properties of the refractories due to hydration. So the presence of lime in magnesia refractories is important, and amount of the same is to be controlled as per the amount of silica present as an impurity. Table 8.1 details the different ratios of lime:silica, and the possible compound formation in the matrix phase and their melting/fusion points. Also, Figure 8.1 shows the CaO–SiO$_2$ phase diagram, detailing the compounds present in the system and the liquid-phase formation.

It can be seen from the Table 8.1 that the low-melting compounds, namely monticellite and merwinite (with melting points between 1500°C and 1550°C), form when the ratio is below 2. Also, these two compounds, if present together, can produce liquid phase even at further lower temperature ~1450°C. Again, when the ratio is above 2, the compounds formed among the impurities in matrix phase have melting points above 2000°C. So, there will be no low-melting liquid-phase formation within the

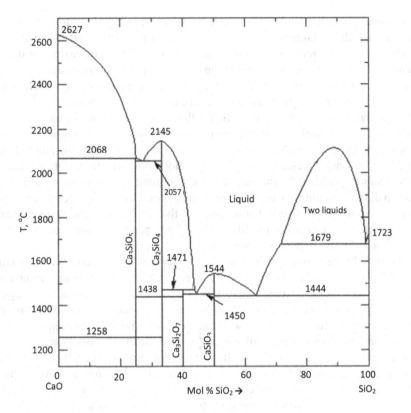

FIGURE 8.1 Phase diagram of lime (CaO)–silica (SiO₂) system.

refractory and, no deterioration in properties of the refractory at high temperatures. Hence, a lime–silica ratio above 2 is necessary for avoiding any formation of these low-melting compounds.

8.5 CLASSIFICATIONS AND PROPERTIES

Magnesia refractories have periclase as the main constituent, with a few compounds based on the impurities present as minor phase, namely, belite ($2CaO\ SiO_2$), alite ($3CaO\ SiO_2$), forsterite ($2MgO\ SiO_2$), monticellite ($CaO.MgO.SiO_2$), and some magnesia ferrite. Depending upon the purity and impurities type and quantity, the magnesia refractories are classified. Mostly, the classification is based on the purity (% of MgO). Higher the purity, better will be properties from corrosion and high-temperature points of view. Depending on the purity, impurity types, and processing conditions, density and porosity values of magnesia refractories vary. Bulk density varies between 2.8 and 3.1 g/cc, and apparent porosity is in the range of 14%–20%. As sintering is difficult for magnesia, generally the strength values are not very high, and varies between 400 and 800 kg/cm².

Though magnesia is a very high-melting oxide, its strength at high temperature is not remarkable. Magnesia has a wide gap between its RUL and PCE values, in comparison to other refractories, say silica or alumina. The main reason for its low-strength values at high temperatures is the presence of the impurities and the formation of low-melting compounds like monticellite, merwinite, and magnesium borate (for sea water magnesia). Also, as these compounds are not highly siliceous in nature, they are low, viscous, and so form liquid flow easily under load and deform the shape easily under load. Hence, thermomechanical properties are not very good, as compared to PCE values. Also, there is a reason from crystallographic point of view too. Magnesia has a face-centered cubic crystallographic structure. Hence, it has the densest atomic packing. So the number of slip systems (combination of slip planes and slip directions) are very high for these structures. Hence, slip occurs very easily in these systems under stressed conditions, as the gap between the two slip planes are the minimum. Generally, the RUL value of magnesia refractory varies between 1550°C and 1650°C, depending on its purity.

Magnesia being a strong basic material, so it has a very good corrosion resistance against basic slags and environments. But, it is highly vulnerable in any acidic environment. It also shows a very strong resistance against high FeO-containing slags, as magnesia can absorb FeO in its structure by forming a solid solution. Slag resistance of magnesia refractories increases with the increase in purity. However, increase in porosity and presence of rough surface causes increase in slag corrosion.

Magnesia has a high thermal expansion properties and has a coefficient of linear thermal expansion of $\sim 13 \times 10^{-6}\,°C^{-1}$, which is very high among the oxides. Due to this expansion properties, magnesia refractories have poor thermal spalling resistance and disintegrate under thermal cycling. Even it is hard to attain 20 thermal cycles (of 1000°C heating and air cooling) for a pure-magnesia refractory. It has been reported that (i) prolonged heating at the highest temperature to form the maximum extent of periclase and to complete any reaction within the impurities, (ii) reduction in FeO content in the composition, and (iii) use of fine grinding and high pressing pressure resulting in greater sintering, improve the thermal spalling character. Also to improve thermal spalling character, alumina or chromia is added to form the respective spinels and thus improves the property.

8.6 EFFECT OF IMPURITIES WITH BINARY AND TERNARY PHASE DIAGRAMS

It has been discussed that magnesia is associated with certain impurities. Natural magnesite source is mainly associated with impurities like CaO, Fe_2O_3 and SiO_2, and sea water source is associated with CaO and harmful B_2O_3. Also, to improve the sintering of strongly ionic-bonded magnesia, mill scale (mainly Fe_2O_3) is added. Hence, the most common impurities present in magnesia refractories are CaO and Fe_2O_3. Now Figure 8.2 shows the binary phase diagram of MgO with CaO and Fe_2O_3. It can be seen that the MgO–CaO diagram has no liquid phase below 2370°C; hence, the presence of only CaO is not at all serious danger for magnesia refractories. Also, a good amount of CaO also goes into the MgO structure, forming a solid solution. So,

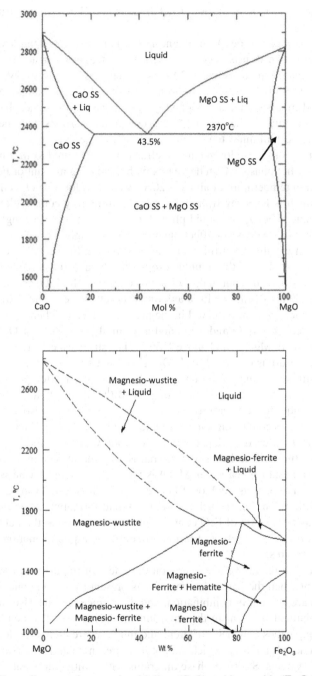

FIGURE 8.2 Phase diagram of magnesia with lime (CaO) and iron oxide (Fe₂O₃).

from the refractory point of view, CaO alone is not a harmful impurity in magnesia refractories.

Again, presence of Fe_2O_3 in magnesia forms magnesiowustite [(Mg, Fe)O] solid solution, in which iron is accommodated in the magnesia structure. Absorption of iron is the maximum at around 1720°C and decreases sharply at lower temperatures. Below about 1000°C presence of iron form two different compounds: magnesio-wustite and magnesio-ferrite ($MgFe_2O_4$). The presence of liquid phase is observed at low temperature in the very high iron-containing portions, and the melting point of Fe_2O_3 is the minimum temperature of liquid-phase existence. For magnesia-rich region (which is the case for magnesia refractories), the liquid forms on melting of the magnesiowustite phase, which decreases with increasing amount of Fe_2O_3. However, the minimum temperature of about 1720°C is reached for a Fe_2O_3 content more than 65 wt%, which is far away from any magnesia refractory composition. Hence, for a lower amount of Fe_2O_3, the liquid-phase formation occurs at very high temperatures, and only iron oxide does not affect the properties strongly.

The effect of other harmful impurities is shown in Figure 8.3. It can be seen that in the presence of SiO_2 the liquidus drops sharply, and the eutectic forms at about 1840°C. But silica does not make any solid solution with MgO, and so even a minute amount of silica will cause a little amount of liquid phase at and above 1840°C. The similar character is also observed in the presence of TiO_2. There is no solid solution formation between MgO and TiO_2, and any small presence of TiO_2 starts forming liquid phase with MgO at and above 1756°C. The strongest effect in reducing the liquidus temperature is observed for B_2O_3. It does not enter into the structure of MgO, and any minute amount of boron can reduce the liquidus temperature to 1366°C. Formation of any Mg–borate phase, like, MgB_4O_7, $Mg_2B_2O_5$, and $Mg_3B_2O_6$, further reduces the liquidus temperature and affects the refractory drastically. Hence, magnesia refractories containing even a minute amount of B_2O_3 will have a little amount of liquid phase even from a temperature of 1366°C. Hence, the hot strength above this temperature will fall drastically. That is why the maximum allowable limit of B_2O_3 in magnesia (in sea water MgO) is 0.01 wt%. Again, if iron is present in its bivalent form, that is as FeO, or if FeO comes from slag, it forms a complete solid solution with magnesia. In such a case, the liquid formation temperature decreases with increasing amount of FeO content, and the minimum at the FeO melting point of 1370°C. As FeO is present in minor amounts, the liquid formation occurs only at high temperatures.

However, all the above discussion on the effect of impurities in MgO refractory is important when the particular impurity is present as a single one. But, in reality, impurities are present in combination, and the refractories are also interacting with various chemicals present in the environment during use. The combined effect of all these impurities is widely different and much severe compared to the effect of a single individual impurity, which cannot be represented and understood by the simple binary systems. Study of phase diagrams with multicomponent systems may be more appropriate; however, finding such phase diagrams and their understanding are difficult. Ternary phase diagrams are somewhat closer to the application conditions as the effect of two separate impurities can be studied simultaneously with varying temperature, and the same for magnesia refractories are discussed below.

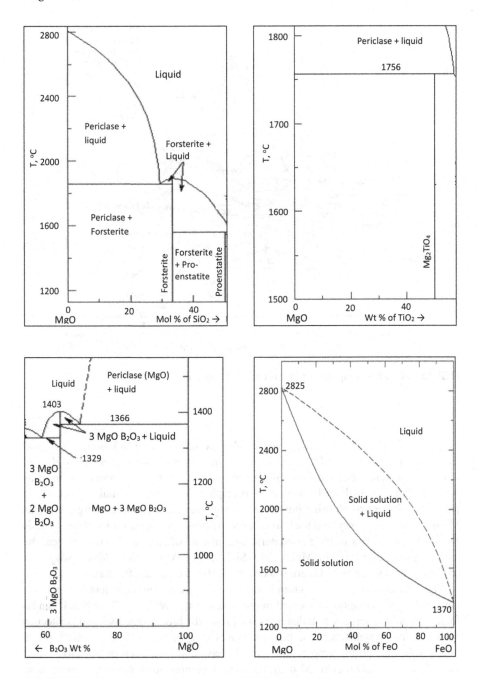

FIGURE 8.3 Phase diagram of magnesia with SiO_2, TiO_2, B_2O_3, and FeO.

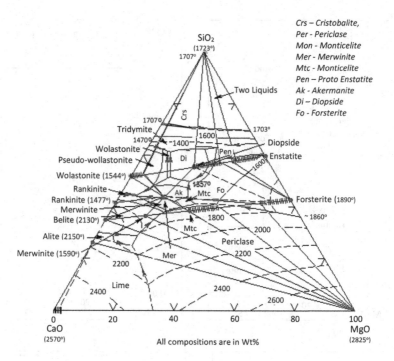

FIGURE 8.4   Phase diagram for MgO–CaO–SiO$_2$ system.

The major impurity oxides present in MgO refractory are CaO, SiO$_2$, Al$_2$O$_3$ and iron oxide (as FeO or Fe$_2$O$_3$). CaO is very common with the sources of MgO and may be associated as lime stone along with natural magnesite sources or used as hydrated lime or hydrated doloma (as precipitating agent) for making sea water magnesia. Silica and alumina impurities in MgO refractories are coming mainly as Mg–silicates and clayey materials associated with the magnesia source. Iron oxide is present as impurities in the raw materials mainly as silicates or ferrites. Hence, the ternary phase diagrams of MgO–CaO–SiO$_2$, MgO–Al$_2$O$_3$–SiO$_2$, MgO–SiO$_2$–iron oxide and MgO–CaO–Al$_2$O$_3$ are important for the magnesia refractories.

MgO–CaO–SiO$_2$ phase diagram is shown in Figure 8.4, and the details of the ternary and binary compounds formed in the system are provided in Table 8.2. It can be found that all the ternary liquidus lines start from the binary eutectics and peritectics present in the systems, and the liquidus moves away from the MgO primary phase region with decreasing temperature. Liquid phase can exist even below 1400°C but away from the MgO region. Also, all the ternary compounds formed are low melting and are far away from the MgO region. The lowest temperature liquid phase observed near MgO-phase region is around diopside (pigeonite) and akermanite. The presence (or formation) of these compounds, even in minute quantity, in MgO refractory confirm the presence of liquid phase, even at 1360°C. Hence purity of MgO used for making the refractory is a critical parameter for the properties and performance of the refractory.

TABLE 8.2

Compounds Formed in the System MgO–CaO–SiO$_2$

Formula	Name	Melting Point (°C)	Melting Character
	Ternary Compound		
2CaO.MgO.2SiO$_2$	Akermanite (Ak)	1454	Congruent melting
3CaO.MgO.2SiO$_2$	Merwinite (Mer)	1590	Incongruent melting
CaO.MgO.SiO$_2$	Monticelite (Mtc)	1503	Congruent melting
CaO.MgO.2SiO$_2$	Pigeonite; (Pgt)/Diopside	1390	Congruent melting
	Binary Compound		
3CaO.SiO$_2$,	Alite or Hatrurite (Hat)	2150	Incongruent melting
2CaO.SiO$_2$	Belite	2130	Congruent melting
3CaO.2SiO$_2$	Rankinite	1475, Dissociation	Incongruent melting
CaO.SiO$_2$	Wollastonite	1540	Congruent melting
MgSiO$_3$	Orthopyroxene (Opx)/ Protopyroxene (Ppx)	1557	Incongruent melting
Mg$_2$SiO$_4$	Forsterite (Fo)	1888	Congruent melting

TABLE 8.3

Compounds Formed in the System MgO–Al$_2$O$_3$–SiO$_2$

Formula	Name	Melting Point (°C)	Melting Character
	Ternary Compound		
2MgO·2Al$_2$O$_3$·5SiO$_2$	Cordierite	Dissociates at 1345	Incongruent melting
4MgO·5Al$_2$O$_3$·2SiO$_2$	Sapphirine		Incongruent melting
	Binary Compound		
MgSiO$_3$	Orthopyroxene (Opx)/Protopyroxene (Ppx)	1557	Incongruent melting
Mg$_2$SiO$_4$	Forsterite (Fo)	1888	Congruent melting
MgAl$_2$O$_4$	Spinel	2135	Congruent melting
2Al$_2$O$_3$.2SiO$_2$	Mullite	1810, Dissociation	Incongruent melting

The ternary system of MgO–Al$_2$O$_3$–SiO$_2$ is shown in Figure 8.5, and the details of the ternary and binary compounds formed in the system are provided in Table 8.3. The phase diagram shows that the ternary liquidus lines starting from the binary invariant (eutectic and peritectic) points at the edge of the ternary diagram and moves to the interior of the system, away from primary phase region of MgO, with decreasing temperature. Two different ternary liquidus lines meet and form a ternary invariant point, and further moves toward final ternary eutectic composition, where the liquid phase exists at the lowest temperature. The lowest liquidus temperature for the MgO–Al$_2$O$_3$–SiO$_2$ system is at the tridymite–protoenstatite–cordierite eutectic point with a temperature of 1345°C, and the next lowest invariant point is another ternary eutectic between the cordierite–enstatite–forsterite system at 1360°C. Both

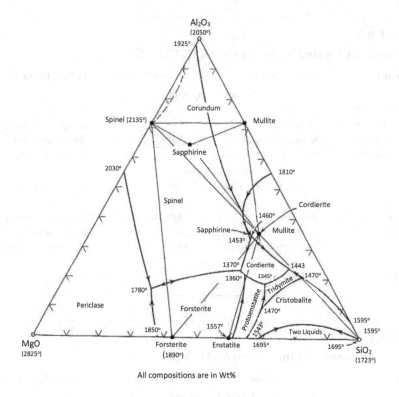

All compositions are in Wt%

FIGURE 8.5 Phase diagram for MgO–Al₂O₃–SiO₂ system.

the lowest eutectic compositions (containing < 30% MgO) are not possible to present in MgO refractories. However, presence of higher amount of impurities and non-equilibrium situations may lead to zonal/local presence of low-melting liquid phases close to the impurity particles, and can result in drastic deterioration in the properties and performances of the MgO refractory.

MgO–SiO₂–FeO system does not show (Figure 8.6) formation of any ternary compound, and a complete solid solution is found between the two silicates – forsterite (Mg₂SiO₄) and fayalite (Fe₂SiO₄) and also between the two end members – MgO and FeO. The details of the binary compounds formed are provided in Table 8.4. There is also one ternary liquidus line that starts at the eutectic (~1860°C) between MgO and Mg₂SiO₄, and the liquidus moves with decreasing temperature toward the eutectic (~1177°C) between FeO and Fe₂SiO₄. This liquidus line has no connection with any other liquidus and is not associated with any ternary peritectic, eutectic, or compound formation. Also, there are two more ternary liquidus lines that start from the MgO–SiO₂ binary system, one from the eutectic point (~1543°C) between MgSiO₃ and MgO and another from the pertectic (dissociation of MgSiO₃) point of ~1557°C. Both the liquidus moves toward the FeO–SiO₂ system with decreasing temperature, and finally meet at the binary eutectic point (~1178°C) between Fe₂SiO₄ and SiO₂.

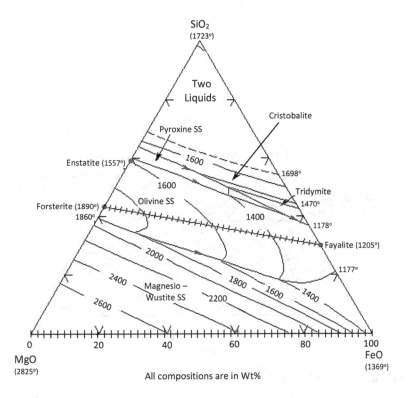

FIGURE 8.6 Phase diagram for MgO–SiO₂–FeO system.

TABLE 8.4
Compounds Formed in the System MgO–SiO₂–FeO

Formula	Name	Melting Point (°C)	Melting Character
	Binary Compound		
$MgSiO_3$	Orthopyroxene (Opx)/ Protopyroxene (Ppx)	1557	Incongruent melting
Mg_2SiO_4	Forsterite (Fo)	1888	Congruent melting
Fe_2SiO_4	Fayalite (Fo)	1205	Congruent melting

These two ternary liquidus lines are also not involved in any compound formation. All the liquidus lines are far away from the primary phase region of MgO, and so they do not have much significance on the properties and performances of the commercial MgO refractories.

The ternary phase diagram of MgO–CaO–Al₂O₃ shows (Figure 8.7) the formation of only one ternary compound 3CaO.MgO.2Al₂O₃ and few binary compounds, as compiled in Table 8.5. There are a number of ternary liquidus lines that start from

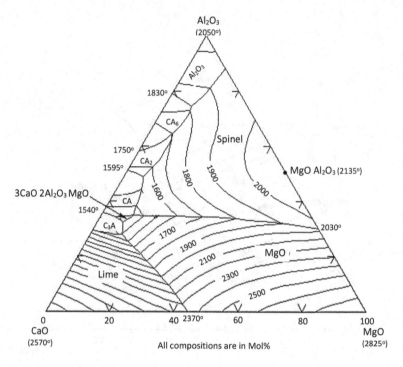

FIGURE 8.7 Phase diagram for MgO–CaO–Al₂O₃ system.

TABLE 8.5
Compounds Formed in the System MgO–CaO–Al₂O₃

Formula	Name	Melting Point (°C)	Melting Character
		Ternary Compound	
$3CaO.MgO.2Al_2O_3$		Dissociates at 1535	Incongruent melting
		Binary Compound	
$MgAl_2O_4$	Spinel	2135	Congruent melting
$3CaO.Al_2O_3$		Dissociates at 1535	Incongruent melting
$5CaO.3Al_2O_3$		1455	Congruent melting
$CaO.Al_2O_3$		1600	Congruent melting
$3CaO.5Al_2O_3$		1720	Congruent melting
$CaO.2Al_2O_3$	Grossite	1789	Congruent melting
$CaO.6Al_2O_3$	Hibonite	1870	Incongruent melting

---

the invariant points (eutectic and peritectic) of the binary systems and move down with decreasing temperature toward the ternary compound present away from the region of MgO. The existence of the lowest temperature liquid phase on the system is ~1500°C, and the same is close to the CaO–Al₂O₃ binary system, exactly opposite to the MgO end of the ternary phase diagram. The ternary compound has also a low MgO content (17 mol% or <10 wt%), and so this low-melting liquid phase and the low-temperature liquidus are far away from the MgO primary region, and are not significant for the MgO refractory.

8.7   MAIN APPLICATION AREAS

The most advantageous property for magnesia refractories is its excellent resistance against basic environment, and so it is highly demanded in applications where refractories have to perform against basic slags and environments. But there are also some serious disadvantages associated with magnesia, like relatively lower hot strength properties, low resistance to thermal shocks, higher thermal expansion and conductivity, etc., have restricted its wide use as a single-component pure-magnesia refractory.

High pure-magnesia refractories (MgO ~ 95%–97%, CaO ~ 2-1.5 and SiO_2 ~ 1%– 0.5%) are used in chemical industries where high basic environment is prevailed. These high pure-magnesia refractories are also important for the upper part of the glass tank furnace regenerators that are prone to be attacked by the alkalis of the batch materials, coming out with the flue gas and get deposited on the top part of checker work. Resistance to alkali and sulfur attack is the prime importance in that region. Incorporation of zirconia (~10%–12%) is also done to improve the abrasion resistance of the regenerator refractories, as solid batch materials are coming out at a high speed with the flue gas and abrade the refractory surfaces.

The next-level purity class of magnesia refractories (MgO ~95%, CaO ~2%, and SiO_2 ~2%) are used in a lime kiln, where lime stone is calcined, and high basic environments prevail. These refractories are also important for backup lining of the electric arc furnaces (EAF). 92% pure refractories with about 2% CaO and ~4% SiO_2 are used for the hot metal mixer in the iron and steel industries. Backup of the lining of converters, ladles, and electric arc furnaces of steel industries are done by using relatively less pure-magnesia refractories (MgO ~87%, CaO ~2%, and SiO_2 ~6%). Relatively inferior magnesia refractories containing about 50% MgO are used for the reheating furnace hearth applications.

SUMMARY OF THE CHAPTER

Magnesia is the most important basic oxide mainly used as basic refractory. Though lime has higher basicity, but due to hydration tendency it is not widely practiced commercially.

There are three major raw material sources of magnesia, namely natural magnesia, sea water magnesia, and brine magnesia. Natural magnesite is mainly associated with impurities like lime, silica, and iron oxide. For sea water magnesia, the most harmful impurity is the B_2O_3. For brine magnesia also, major impurities are iron oxide and silica. There is also fused magnesia available, produced by electro-fusion route and used to improve the corrosion resistances.

Conventional shaped refractory manufacturing technique is used for making magnesia refractories, but as magnesia is a strong ionic-bonded oxide, its sintering requires higher temperature, and mill scale (mainly Fe_2O_3) is added as a sintering aid during mixing.

Among impurities, silica is very common in natural magnesite, which can form low-melting impurity phase (silicates) that can strongly deteriorate the hot properties. Hence, it is practiced to have high lime content in magnesia refractory having silica as an impurity. The presence of lime will react with silica, forming high-temperature melting di-/tri-calcium silicates, thus improving the hot properties. So lime-to-silica ratio is important and is maintained above 2.

Magnesia has relatively higher thermal expansion property, resulting in poor thermal shock resistances. It has very high corrosion resistances against basic slags/environments, but for slags with high FeO, it may form a solid solution and get dissolved in slag causing high wear out of the refractory.

It is mostly used in the backup lining of steel converters, steel ladles, electric arc furnaces, etc.; and top part of regenerators of glass-melting tanks, in lime kilns, etc.

QUESTIONS AND ASSIGNMENTS

1. Why is lime not commonly used as basic refractory?
2. What are the different raw materials sources for magnesia refractories? Discuss them.
3. What are the main impurities present in different raw material sources for magnesia? How they affect the quality of magnesia?
4. How lime–silica ratio affects the properties of magnesia refractories?
5. Discuss the manufacturing technique in detail for making magnesia refractories.
6. Discuss why magnesia has much lower thermomechanical properties compared to its PCE.
7. Why CaO is not a harmful impurity for magnesia but SiO_2 and B_2O_3 are?
8. Discuss the main application areas of magnesia refractories.
9. Describe with phase diagram how a magnesia refractory will behave with a slag containing high FeO.

BIBLIOGRAPHY

1. J. H. Chesters, *Refractories- Production and Properties*, Woodhead Publishing Ltd, Cambridge, 2006.
2. P. P. Budnikov, *The Technology of Ceramics and Refractories*, Translated by Scripta Technica, Edward Arnold, The MIT Press, Massachusetts, US, 4th Ed, 2003.
3. A. R. Chesti, *Refractories: Manufacture, Properties and Applications*, Prentice-Hall of India, New Delhi, India, 1986.
4. *Refractories Handbook*, The Technical Association of Refractories, Tokyo, 1998.
5. *Harbison Walker Handbook of Refractory Practice*, New York, 2005.
6. Y. Yin and B. B. Argent, The phase diagrams and thermodynamics of the ZrO_2-CaO-MgO and MgO-CaO systems. *Journal of Phase Equilibria*, 14 [5] 588–600 (1993).
7. E. Woermann, B. Brezny and A. Muan, Phase equilibria in the system MgO–iron oxide–TiO_2 in air. *American Journal of Science*, 267A, 463–479 (1969).

8. S. Miyagawa, S. Hirano and S. Somiya, Phase relations in the system MgO-B2O3 and effects of boric oxide on grain growth of magnesia. *Yogyo Kyokaishi*, 80 [2] 53–63 (1972).

9. P. Wu, G. Eriksson, A. D. Pelton and M. Blander, Prediction of the thermodynamic properties and phase diagrams of silicate systems–evaluation of the FeO-MgO-SiO$_2$ system. *ISIJ International*, 33 [1] 26–35 (1993).

10. V. Swamy, S. K. Saxena and B. Sundman, An assessment of the one-bar liquidus phase relations in the MgO SiO$_2$ system. *Calphad: Computer Coupling of Phase Diagrams and Thermochemistry*, 18 [2] 157–164 (1994).

11. B. Phillips, S. Somiya and A. Muan, Melting relations of magnesium oxide-iron oxide mixtures in air. *Journal of the American Ceramic Society*, 44 [4] 167–169 (1961).

12. J. D. Panda, *Proceedings of Indian National Science Academy*, 50A [5] 428–38 (1984).

13. A. S. Bhatti, D. Dollimore and A. Dyer, Magnesia from seawater: A review. *Clay Minerals*, 19, 865–875 (1984).

14. H.M. Richardson, M. Lester, F.T. Palin and P.T.A. Hodson, The effect of boric oxide on some properties of magnesia. *Transactions of the British Ceramic Society*, 68 29–31 (1969).

15. W. M. Huang, M. Hillert and X. Z. Wang, Thermodynamic assessment of the CaO-MgO-SiO$_2$ system, *Metallurgical and Materials Transactions A*, 26A [9] 2293–2310 (1995).

16. M. L. Keith and J. F. Schairer, The stability field of sapphirine in the system MgO - Al$_2$O$_3$- SiO$_2$, *The Journal of Geology*, 60 [2] 182 (1952).

17. P. Wu, G. Eriksson, A. D. Pelton and M. Blander, Prediction of the thermodynamic properties and phase diagrams of silicate systems–evaluation of the FeO-MgO-SiO$_2$ system, *ISIJ International*, 33 [1] 26–35 (1993).

18. M. T. Mel'nik, A. A. Kachura and N. V. Mokritskaya, Fusion diagram for the pseudo-system CaO Al$_2$O$_3$- MgO·Al$_2$O$_3$- MgO, *Refractories*, 30 227–229 (1989).

19. S. Behera and R. Sarkar, Sintering of magnesia: effect of additives, *Bulletin of Materials Science*, 38 [6] 1499–1505 (2015).

9 Dolomite Refractories

9.1 INTRODUCTION

From the basicity point of view, lime has an edge over magnesia, but hydration has restricted its success. However, a mixture of lime and magnesia can be used having both the positive and negatives of both the oxides. Hydration tendency of lime is thus reduced with increase in basicity for magnesia. Such a natural raw material source is dolomite.

Dolomite is a double carbonate of calcium and magnesium or calcium magnesium carbonate [$CaMg(CO_3)_2$], occurring naturally in many countries and widely used in many industrial applications. The mineral was first identified by Count Dolomien in 1791 and named after its discoverer. Being a natural mineral, usually dolomite contains impurities, chiefly silica, alumina and iron oxide. For commercial purposes, the percentage of combined impurities should not go beyond 7%, above which it becomes unsuitable for any industrial use. High-impurity-containing dolomite is used only for road ballasts, building stones, flooring chips, etc. For refractory industries, the purer variety of dolomite is used with the minimum MgO content of 20 wt% and combined CaO + MgO min 52 wt%. Dolomite with lower purity are important for other industries, like iron and steel, ferromanganese, glass, fertilizer, etc. Table 9.1 gives an idea about the quality of the raw dolomite and their recommended use.

In iron and steel industry, dolomite is important as a flux material. Dolomite used in blast furnace, sinter, and pellet plants are of relatively inferior quality (MgO min 19% and combined CaO + MgO min 46%), as compared to that used in steel-melting shop (MgO min 20% and combined CaO + MgO min 50%). Dolomite is important in iron-making and steel-making processes to produce a highly basic low-melting liquid phase that can extract out the gangue (impurity) materials present (silica, alumina etc.) in iron ore and form the slag easily. For ferromanganese industry also the dolomite used are of similar quality to steel-grade ones (MgO 19%–20%, CaO 28%–30% SiO_2 2%–5%, and Al_2O_3 2%–2.5%). But for ferromanganese industries, dolomite should be hard and fine grained, because crystalline dolomite gives fritting effect in the furnace. In the glass industry, dolomite is a source of lime, essentially required for the chemical property of glass (corrosion resistant) and also as a source of MgO that helps in preventing crystallization. Also, dolomite provides low-temperature eutectics (due to presence of two basic oxides) with acidic systems (silica in glass) and helps in glass formation. But it must be nearly free from iron or any other coloring oxides. Dolomite is used to fertilize industries mainly for soil amendments, and must have $MgCO_3$ + $CaCO_3$ min 90% and finer than 2 mm size.

9.2 RAW MATERIALS AND SOURCES

Dolomite is a naturally occurring sedimentary and metamorphic mineral, found as the principal mineral in widely and abundantly available dolostones and meta-dolostones.

DOI: 10.1201/9781003227854-9

TABLE 9.1

Quality of Raw Dolomite and Recommended Industry for Application

Recommended for	Refractory Industry	Steel Making	Iron Making
$(CaO + MgO)$ Content, % mass, *min*	52	50	46
CaO Content, % mass, *min*	–	30	–
MgO Content, % mass, *min*	20	20	19
SiO_2 Content, % mass, max	2	2	–
Total acid insoluble, % mass, max	2	3	8
Alkali content, % mass, max	0.2	0.2	0.2

It is also an important mineral in certain limestones and marbles, where calcite is the principal mineral present. Dolomite is also found as a hydrothermal vein mineral, forming crystals in cavities, and in serpentinites and similar rocks. Theoretically, dolomite contains 54.35% $CaCO_3$ and 45.65% $MgCO_3$, and as per oxide contents it has 30.4% CaO, 21.7% MgO and 47.9% CO_2. In nature, considerable variations in the composition of dolomite relating to lime and magnesia percentages are found. When the percentage of $CaCO_3$ increases by 10% or more than the theoretical composition, the mineral is termed "calcitic dolomite," "high-calcium dolomite," or "lime-dolomite." With further decrease in percentage of $MgCO_3$, it is called "dolomitic limestone." When the $MgCO_3$ content is between 5% and 10%, it is called "magnesian limestone," and when it is up to 5%, then it is considered as limestone for all purposes in trade and commercial parlance.

As a natural source, dolomite is available as massive layers in dolostones that are very common in ancient rocks. Dolomite is rarely found in modern sedimentary environments but the rock record. They can be geographically extensive and even up to few hundred meters thick. Most of the rocks that are rich in dolomite were originally deposited as calcium carbonate muds and post-depositionally altered by magnesium-rich pore water to form dolomite. Over time, built-up pressure transformed the mixed composition into dolomite. Eventually, the pressure becomes so great that the mixture hardens into a rock with a high amount of foliation. There is also a high magnesium content to the rock, which is due to magnesium replacement for limestone.

Although rocks containing dolomite are found all over the world, the most important availability of dolomite are located in the Midwestern United States; Ontario, Canada; Switzerland; Pamplona, Spain; Mexico, etc.

9.3   BRIEF OF MANUFACTURING TECHNIQUES

Dolomite is a double carbonate that on heating decomposes completely above 900°C. The product resulting from this relatively low-temperature calcination is highly porous and reactive, and is known as "calcinated dolomite" or "doloma."

$$CaMg(CO_3)_2 = CaO.MgO + 2CO_2$$

This calcined dolomite (or doloma) is also fired for dead burning at high temperatures, and contains theoretically 58% of CaO and about 42% of MgO. The presence of free CaO absorbs moisture from the environment and causes hydration.

$$CaO + H_2O = Ca(OH)_2$$

This hydration reaction is associated with volumetric expansion that causes the materials to break down to powder form (hydration/slaking). Hence, any refractory manufactured from the material will also undergo the same hydration process and will shatter. Thus, the doloma is unstable even after dead burning. This hydration may be restricted by converting the lime part of dolomite into an inactive silicate phase or some other compounds having a high softening point too. Silica is added to convert the CaO into calcium silicates. Monocalcium silicate is avoided for its low softening point (~1540°C), but di- and tri-calcium silicates are allowed to form and preferred in the composition (due to their high-temperature softening character) by reacting all the CaO in the dolomite composition with silica (present in dolomite and added separately). Conversion of CaO to high-temperature calcium silicates (mainly di-calcium silicate, preferred due to its high melting point) prevents any hydration, and the final doloma is nonhydratable and stable. This type of calcined product is termed as "stabilized dolomite," and the process is called stabilization.

But di-calcium silicates have different crystallographic/polymorphic forms that change during heating and cooling, and due to the great difference in the specific gravity of these polymorphic phases, there are huge volumetric changes that cause shattering of the shapes. Phase inversion of di-calcium silicate from its beta to gamma phase during cooling at ~675°C results in a volume expansion of ~10%. This expansion is sudden during cooling, and causes cracking and shattering of the refractory. To avoid such disintegration Fe_2O_3 is added that restricts the formation of gamma phase and stabilizes the beta phase of di-calcium silicate. The addition of boron and phosphates also helps in stabilization of dolomite. But due to the presence high amount of impurity phases like silica, iron oxide, etc. low-melting compounds form in doloma refractory due to the reaction between the impurities and CaO and MgO of doloma. Hence, the stabilization of dolomite is good for hydration resistance but parallelly results in poor hot properties, especially the hot strength.

To reduce this reduction in high-temperature properties, the concept of stabilization is modified to "partial stabilization" in which much lower amount of silica is added to react with CaO present on the surface of doloma just to convert the surface into di-/tri-calcium silicates. As hydration is a surface activity, the coating of silicates on the surface protects the material from hydration. Again, as only surface CaO is converted to silicate, amount of silica required for such conversion and the amount of low-melting liquid phase that may form are much lower than that of the stabilized dolomite. So the high-temperature properties of dolomite are not strongly affected. Instead of addition of silica, serpentine ($3MgO\ 2SiO_2\ 2H_2O$) is also used to form high-temperature silicates producing excess MgO in the composition.

$$6(CaO.MgO) + 3(MgO\ 2SiO_2\ 2H2O) = 2(3CaO.SiO_2) + 9MgO + 2H_2O$$

Calcining and dead burning (with additions for stabilizers like silica) of dolomite is done in a rotary kiln or shaft kiln. The dolomite stone is crushed as per the kiln type and size, and then washed to make it free from impurity particles, like clay. The total treatment process to get the sintered doloma grains may be divided into four steps. They are drying (to remove external and any internal moisture), calcination (to remove the gaseous material like CO_2), firing or dead burning (to get dense sintered aggregates with reduced porosity, final temperature depends on the composition and desired properties in the product), and cooling.

As the process of stabilization basically imparts impurity, and a comprise is required between hydration tendency and high-temperature properties, new techniques have come up to produce very high-fired doloma aggregates to reduce hydration tendency, and processing them fast to convert to refractory bricks and cover them to avoid any hydration. In one technique, termed as single-pass process, where the dolomite is passed through the kiln once only, involves very high-temperature dead burning of the natural purer dolomite, above 1800°C, to make the aggregate immune and increase the grain size to reduce the hydration tendency. Also, processing the aggregates to make bricks in a fast process (brick manufacturer itself makes the single-pass dolomite aggregate in the same site to avoid any time delay for transportation and reduce the chances of hydration). In another route, termed as double-pass process, a two-stage firing is involved with a palletization process in between. In the first firing, dolomite is calcined to produce active oxide, which is pelletized and fired at high temperatures to immune the material. These firing temperatures are above conventional dead burning temperatures. The double firing process is obviously energy intensive and makes the product costly. But the palletization process involved in between helps in uniformity of densification, and growth in the doloma aggregates with improved and uniform final sintered properties. Since the impurity (like silica, alumina, iron oxide) present is very low, and the low-melting liquid phases formed are minumum in amount, obtaining high densification is sometimes difficult in the dead burning process. Temperatures more than 1850°C are usually required to achieve satisfactory density and hydration resistance.

Generally high-grade dolomite, containing combined impurities less than 3%, is selected for dead burning. As it is difficult to densify high-purity dolomite in a rotary kiln, it is customary to use some mineralizers to facilitate dead burning and get well dense aggregates. Iron oxide is a common additive for this purpose. The manufacturing process parameters vary with the grade of dead-burnt dolomite to be produced. The dolomite after dead burning is cooled in either rotary or reciprocating recuperative coolers.

Recently, there are trends to use synthetic dolomite clinkers that contain no natural dolomite. Here, calcium hydroxide is mixed with magnesium hydroxide with a small amount of additives like iron oxide and fired at high temperature to get dead, burnt dolomite clinker. The advantages of this type of aggregates are that the impurity phase present is very minimum and completely controlled, and the composition can be adjusted as per the requirement in a particular application (CaO-/MgO-rich compositions). Hence, the properties can be tailored, and generally performances of the synthetic aggregates are better compared to the natural ones, especially for the hot strength and corrosion-resistance points of view.

Clinker/aggregates obtained by the dead burning process is crushed and ground to desired particle size fractions, and mixed in proportions to obtain optimum packing. The fractions are first dry mixed in a mixer, and then ~4% water is added for hydraulic pressing of the stabilized dolomite. Mix batch is then hydraulically pressed at 80–120 MPa and then dried. For water-containing compositions, drying is critical, as cracks may appear from hydration. To avoid any crack formation during drying, it is done with a stream of dry and clean air at temperatures below 60°C. To eliminate the risk of hydration, in many a case, water is avoided and wax is used. To get a good flowability of wax and uniformity in the mixture, little heating (~60°C) is done during the mixing process. The firing of dolomite refractories is done in both batch and continuous types of kilns. The temperature of firing the shapes depends on the chemical purity of the dolomite aggregate used. For high pure, dead-burnt (not stabilized) dolomite, firing is done more than 1700°C to get strong and well-densified products.

Other than this conventional sintered dolomite refractories, there is also tar-/pitch-bonded dolomite bricks used in steel-making process. In this case, raw dolomite is directly used to make the refractory without calcination. Here, first raw dolomite is crushed and ground, and the different fractions are mixed with ~10% tar/pitch. Mixed compositions are then pressed hydraulically and tempered below 400°C. Due to low-temperature processing, the cost of these types of refractories is low. Again, the presence of carbon (coming from the fixed carbon of tar/pitch) in these refractories helps in increasing the corrosion resistance.

9.4 CLASSIFICATIONS AND PROPERTIES

Dolomite refractories can be of different types as per the starting material used. It can be fully stabilized refractory, which has relatively high silica (impurities) content, having strong hydration resistance, but at the cost of formation of the liquid phase at lower temperatures and poor hot strength properties. Partially stabilized dolomite refractories are having lower extent impurities and have only low-temperature liquid coating at the surface, resulting in strong hydration resistance and little improved hot properties compared to fully stabilized ones. Refractories with high-temperature dead-burnt dolomite clinkers with the minimum impurities show relatively better properties, but continuous attention and precautions are required to avoid hydration of the product.

Carbon-bonded doloma refractories are also a special class that is economic and hydration resistant. These unfired dolomite refractories can be further subdivided according to the type of carbon binder used. Tar- or pitch-bonded unfired dolomite refractories are thermoplastic in nature, but resin-bonded ones are thermosetting in character. Again pitch-bonding graphitizes at higher temperatures resulting in better oxidation resistance compared to that of the "glassy natured carbon" from the resin-bonded ones. But the resin-bonded ones show higher strength values and lower fume emissions during preheating.

In general, properties of doloma refractories vary widely depending on the types of the aggregates used, their compositions, and process of manufacturing. Doloma is made up of two very stable oxides and that too of high basic character. Hence, the refractory is very strong against molten steel and any basic slag. Generally, in contact with slag that is not saturated with the basic components, a layer of dense and highly viscous Ca/Mg silicate liquid forms due to the reaction between doloma refractory and slag components on the hot face of the refractory. Thus, restricting the further slag corrosion and penetration. Thus, the overall corrosion resistance of the refractory is improved. Further improvement in the slag resistance of doloma refractories is done by addition of high-purity magnesia in it. But, if the slag is highly acidic or contains lesser amount basic components compared to that of R_2O_3 components, low-melting and low-viscous liquids in the $CaO-Al_2O_3-Fe_2O_3-SiO_2$ system form, thus resulting in high corrosion of the refractory. Also, doloma refractories are weak against slags containing high amount of iron oxide.

Softening of these refractories occur at very high temperatures, but the hot strength drastically deteriorates with the amount of impurity phase. Formation of low-melting and low-viscous Ca/Mg silicate phases is mainly responsible for lower strength properties at high temperatures. The presence of any other impurity in the composition further deteriorates the quality. Good-quality doloma has refractoriness (PCE) ~2300°C but RUL value ~1650°C–1700°C.

For doloma refractory, hydration resistance is important due to the presence of lime in it. Any free lime present in doloma reacts with atmospheric moisture and makes the material crumble on hydration. The degree of hydration is dependent on the composition (amount of lime and impurities), densification, surface porosity, time, temperature, and relative humidity. The degradation of doloma on hydration occurs in two stages:

1. the formation of $Ca(OH)_2$ on the surface, and
2. breaking apart of the refractory (grains) into smaller particles due to expansion.

The hydration tendency is a major problem for the doloma refractory manufacturers, as the sintered grains cannot be stored for long. To control the hydration problem, attention and precaution at the critical stages of manufacturing of dead-burnt aggregates and bricks are required. Immediate use of the freshly prepared doloma aggregates in the brick making also must be practiced to reduce the risk of hydration with time. Better packaging techniques and equipment are to be used to retain the dehumidifying conditions for the fired bricks that can further control the crisis to a great extent. During application, the refractories are constantly at high temperatures, and so the chances of hydration (low-temperature activity) are less.

Doloma is a solid solution of two oxides that are having relatively higher thermal conduction and thermal expansion properties among the different refractory oxides. Hence, it conducts higher amount of heat, may cause greater heat loss, and moreover higher expansion properties cause degradation in the thermal shock properties. Again, for fully or partially stabilized dolomites, type of silicate phase formed is important. The presence of di-calcium silicates (C_2S) is good due to its volume stability at high temperature, but phase inversion on cooling is a serious concern and requires proper attention. Again, a significant improvement in the thermal shock resistance of doloma refractories is achieved by the addition of zirconia. This is due to the microcracking effect caused by the formation of calcium zirconate phase (the formation is associated with volumetric expansion).

9.5 EFFECT OF IMPURITIES WITH BINARY AND TERNARY PHASE DIAGRAM

Doloma is a solid solution of CaO and MgO, and the details of the phase diagram are given in Figure 8.2. Both the oxides are individually high melting, and have melting points 2570°C and 2825°C, respectively. Though they satisfy most of the criteria to form a complete solid solution, a greater difference in their ionic sizes does not allow them to do so. They react and form a eutectic at 2370 with a composition of 43.5% MgO and 56.5% CaO. But at room temperature, these oxides exist as a mixture.

Two-component phase diagrams related to MgO are described in Chapter 8. Here, the CaO-related phase diagrams are discussed. The phase diagram of the CaO–SiO_2 system is also detailed in Chapter 8, Figure 8.1. The presence of silica in CaO reduces the liquidus temperature, and liquid phase starts forming at the peritectic temperature of ~2070°C, with a peritectic composition of ~25 mol% SiO_2. Generally, the naturally available doloma does not have silica more than this amount. So the presence of silica does not affect much on the liquid phase formation with CaO. But, in a ternary system of CaO–MgO–SiO_2 the liquid phase may start even at a much lower temperature of ~1350°C close to diopside (CaO.MgO.$2SiO_2$) composition.

Details of the phase diagram in the Ca–Fe–O system is given in Figure 9.1. The figure shows that both CaO and FeO can accommodate each other in their structure to a limited extent forming a partial solid solution, with the maximum limit at the peritectic temperature ~1100°C. But the effect of FeO on CaO is drastic; other than forming solid solution, the formation of liquid phase starts even at the peritectic temperature, so when the amount of FeO exceeds the solid solution limit, liquid phase will exist even above 1100°C. However, at low temperature, FeO will exist in CaO system as di-calcium ferrite ($Ca_2Fe_2O_5$) form, the only compound formed in the CaO–FeO system that dissociates to lime (CaO) solid solution and wustite (FeO) solid solution at ~1100°C.

Again the presence of Fe_2O_3 in CaO does not form any solid solution but drastically reduces the liquidus temperature to 1450°C, which is the eutectic temperature between CaO and the compound di-calcium ferrite ($Ca_2Fe_2O_5$). Hence, this signifies that presence of any amount of Fe_2O_3 in CaO system will start producing liquid phase above 1450°C, and the amount of liquid is directly proportional to the amount of Fe_2O_3 present.

Dolomite is a naturally available double oxide, and the impurity commonly associated with it is clay. Both the oxides (CaO and MgO) of dolomite are basic in nature, and the components of clay, that is SiO_2 and Al_2O_3 are deadly for dolomite, as SiO_2 is highly acidic, and Al_2O_3 shows little acidic character at higher temperatures. Hence, for dolomite refractories the most important ternary phase diagrams are CaO–MgO–SiO_2 and CaO–MgO–Al_2O_3 systems. Both the phase diagrams are described under magnesia refractories in Chapter 8.6, and the diagrams are elaborated as Figures 8.4 and 8.7.

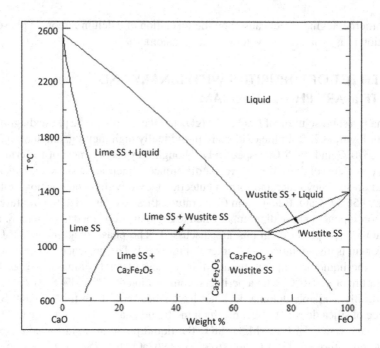

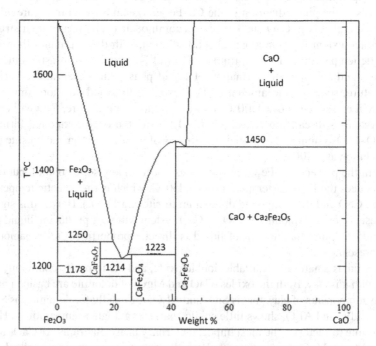

FIGURE 9.1 Phase diagram of lime (CaO) with FeO and Fe_2O_3.

9.6 MAIN APPLICATION AREAS

Dolomite refractories were widely used in steel-making applications till say the 1980s, especially in the converter lining. But as the use of magnesia–carbon refractory has increased, the trend for using dolomite in converter has gone down abruptly. Dolomite was first used as refractory in the linings in open hearth furnaces or Thomas converters in 1878, where sodium silicate-bonded dolomite stone was used and calcined in situ in the converter. Carbon-bonded doloma was also important for the hearth application, as the basic process of steel making gained ground, and importance of removal of phosphorous and sulfur were recognized. During that days, much of the refractory were allowed to be consumed by the steel-making process, and the doloma refractory used was not beneficiated and was also not highly pure.

Magnesia refractories started replacing doloma refractories due to higher refractory character and hydration resistance. Use of doloma refractory as a source of lime in the steel-making process was also no longer required as free lime addition was incorporated in the steel-making process. Carbon-bonded doloma refractories were still continued. In early of the second half of the last century, doloma were still competing with magnesia due to abundant availability and economy, mainly for LD converters and electric arc furnaces. But the later availability of larger grain magnesia and development of magnesia–carbon refractories has slowly and completely removed the use of doloma refractories in primary steel making. But still it is important in secondary steel-making processes.

For iron and steel industries, fired doloma refractories with 60% CaO and 40% MgO content are important for AOD (argon oxygen decarburization) converter and LRF (ladle refining furnace). Also, they are used in steel ladles. Again, refractories with ~40% CaO and ~60% MgO are used for AOD and VOD (vacuum oxygen decarburization) converters. Pitch-bonded doloma with 60% CaO and 40% MgO content are important for steel ladle and LRF applications, whereas similar composition with resin bonding are useful for AOD, VOD, LRF, and steel ladle applications. Dolo–carbon refractories, containing graphite in the compositions with resin bonding, are useful for the impact pads, slag lines, and other high-wear areas in steel converters and steel ladles.

Use of dolomite refractories in burning zones of cement rotary kilns was started around the 1960s, mainly due to chemical compatibility with the cement manufacturing process and easy formation of a protective coating in the burning zone. The resistance of dolomite against alkalis is also important for cement kiln applications. ZrO_2 addition is also done in these bricks to improve the thermal shock resistance. Moreover, zirconia has a high refractoriness, and is abrasion and corrosion resistant to clinker minerals. Zirconia-containing doloma refractories have a high thermal elasticity and excellent thermal shock resistance. Formulations have also been developed using magnesia additions in dolomite refractory to extend its use in the transition zones of the cement rotary kilns. Thus, chrome-free linings in the cement rotary kiln application was possible for environment-friendly manufacturing.

For the cement-making industry, 60% CaO and 40% MgO-containing fired doloma refractories are important for the burning zones. For thermal shock-prone areas of burning zones, ~58% CaO, ~38% MgO, and ~2% ZrO_2-containing compositions are used. High density, low porosity, less permeable brick with similar compositions are important for burning zones prone to chemical attack and gas penetration. Magnesia-enriched doloma compositions (~48% CaO and ~48% MgO) with (~2%) or without ZrO_2 are important for the transition zones of cement kilns and burning zones with the unstable protective coating. Excess MgO in fine fraction helps to form a coat on free lime, and insulates and protects the refractory against the reaction with sulfur-containing volatiles in the kiln atmosphere.

Dolomite is also important as refractory repair mass. Sometimes, it is more suit-able as a repair mass than shaped refractory due to its associated defects and disad-vantages. About 60% CaO, 35% MgO and ~4% Fe_2O_3-containing dolomite fettling mass is important for the wall and bottom repair of steel-melting electric arc fur-naces. Similar composition with lower Fe_2O_3-containing ramming mass is important as the back-filling mass for AOD and LRF applications. Again, ~58% MgO and ~33% CaO-containing composition with gunning consistency is important for hot repair of steel converter, ladle, and electric arc furnaces. Also, ~55%–57% MgO and ~40% CaO-containing compositions are important for cold repair of the hearth of electric arc furnace and AOD vessel,

Recently dolomite refractories have found a newer and technologically critical application for producing of clean steel. Aluminum-killed steel manufacturers typi-cally face a crisis of alumina build-up and clogging of the continuous casting refrac-tories. Alumina clogging affects the life and performance of the refractories, reduces casting rate and productivity, and deteriorates steel quality. This may occur in the ladle shroud, tundish nozzle, tundish slide gate plate bore, submerged entry shroud (SES), and submerged entry nozzle (SEN). Alloying elements like Ti, Ce, La, etc. enhances the build-up and enhances the crisis. Doloma-based refractory composition (with high CaO content, ~60%) in the metal contact areas does not allow the forma-tion of alumina build-up due to the formation of low-melting Ca–aluminate phases with the alumina depositions and prevents the clogging. This improves the overall performance of continuous casting process and quality of steel.

SUMMARY OF THE CHAPTER

Dolomite is a double carbonate of Ca and Mg, and doloma is used as a refractory, which is a solid solution of CaO and MgO.

Doloma, containing about 58% of CaO, is highly hygroscopic in nature, reacts with the moisture present in the atmosphere, and forms calcium hydroxide, associated with a volume expansion, and causing cracking and shattering of the refractory shape.

In earlier days, intentionally low-melting compounds were allowed to form to convert all the CaO to silicate-based compounds, and hydration tendency were pre-vented. These completely converted volume-stable dolomites were termed as stabi-lized dolomite. But the presence of low-melting compounds drastically deteriorated the hot properties.

To improve the hot properties, only the surface of the doloma was allowed to coat with such low-melting compounds, such products were termed as partially stabilized dolomite.

To avoid any incorporation of impurity phases and having hydration resistance, dolomite can be fired at a very high temperature and used immediately for brick making after dead burning using nonaqueous binders. But, after firing of the refractory shape, they need to be specially packed to avoid any contact with the atmospheric moisture.

Doloma refractories are very strong against any basic slag and mostly used for this advantage only. But, other than hydration tendency, it has drawbacks like higher thermal expansion character and lower thermal shock resistances.

Doloma is mostly useful for the secondary steel-making processes and mainly used for the lining of LRF, AOD, VOD, EAF, etc. Other than conventionally shaped refractory, doloma is also important as repair mass, as fettling, ramming, gunning mass in all the different secondary steel-making vessels. Also, doloma refractories are important for burning and transition zones of cement rotary kilns.

QUESTIONS AND ASSIGNMENTS

1. Describe the raw material dolomite.
2. Discuss the advantages and disadvantages of doloma refractories.
3. What is stabilization, and why it is required for doloma?
4. What are the techniques used to restrict the hydration of doloma refractories?
5. What is the role of iron oxide in the stabilization of doloma refractories?
6. Discuss and compare the fully stabilized and partially stabilized doloma.
7. Describe the manufacturing technique of doloma refractories.
8. Detail the properties of doloma refractories.
9. Discuss the effect of impurities on doloma refractories.
10. Discus the main applications of doloma refractories.
11. What advanced applications doloma can have?

BIBLIOGRAPHY

1. Charles Schacht, *Refractories Handbook*, CRC Press, Boca Raton, US, 2004.
2. A. R. Chesti, *Refractories: Manufacture, Properties, and Applications*, Prentice-Hall of India, New Delhi, India, 1986.
3. J. H. Chesters, *Refractories- Production and Properties*, Woodhead Publishing Ltd, New Delhi, India, 2006.
4. P. P. Budnikov, *The Technology of Ceramics and Refractories*, Translated by Scripta Technica, Edward Arnold, The MIT Press, 4th Ed, 2003.
5. *Refractories Handbook*, The Technical Association of Refractories, Tokyo, 1998.
6. *Harbison Walker Handbook of Refractory Practice*, New York, 2005.
7. *Dolomite for Metallurgical and Refractory Use — Specification, Indian Standard Specification, IS-10346: 2004*, Bureau of Indian Standards, New Delhi.
8. M. Rabah and E.M.M. Ewais, Multi-impregnating pitch-bonded Egyptian dolomite refractory brick for application in ladle furnaces. *Ceramics International* 35 813–819 (2009).
9. H. I. Moorkah and M.S. Abolarin, Investigation of the properties of locally available dolomite for refractory applications. *Nigerian Journal of Technology*, 24 [1] (2005).

10. *Product Catalog*, Resco Products, Inc., PA.
11. M. Hillert, M. Selleby and B. Sundman, *An assessment of the Ca-Fe-O system Metallurgical Transactions A*, 21 [10] 2759–2776 (1990).
12. R. E. Johnson and A. Muan, Phase Equilibria in the System CaO-MgO-Iron Oxide at 1500°C. *Journal of the American Ceramic Society*, 48 [7] 359–364 (1965).

10 Chromite and MgO– Cr$_2$O$_3$ Refractories (Chrome–Mag and Mag–Chrome)

10.1 INTRODUCTION

This chapter covers three different types of refractories, but all of them have a common component, chromite. Chromite is the principal ore of chromium in which the transition metal exists as a complex oxide (FeO.Cr$_2$O$_3$) form. Chromite is the only economically extractable natural source of chromium, and in refractory field, it is the prime source for the desired component chromia (Cr$_2$O$_3$).

The usefulness of chromia as a refractory is based on its high melting point of 2180°C (3960°F), moderate thermal expansion, neutral chemical behavior, and relatively high corrosion resistance. Chromite enhances thermal shock and slag resistance, volume stability, and mechanical strength of a refractory. In contact with ferric oxide (Fe$_2$O$_3$), it forms a complete solid solution (a homogeneous crystalline phase composed of Fe$_2$O$_3$ and Cr$_2$O$_3$ dissolved in one another), and in contact with ferrous oxide (FeO), it forms an iron chromate spinel and expands considerably, causing the shape to crumble (bursting).

Again chromium, being a transitional element, changes its valency with oxygen concentration (temperature) and results in different colors. Also hexavalent chromium is a health-hazardous material. Chromium has a higher diffusivity compare to other refractory oxides, and diffuses out from the refractory, enters into other systems, and colors it. Due to these, pure chromite refractory is relatively rare, and it is used in combination with other oxides, mostly magnesia.

Refractory-grade chromite requires high purity (the amount of combined Cr$_2$O$_3$ and Al$_2$O$_3$ should exceed 57%) with very low silica content. Other oxides present in chromite affects the refractoriness of the material and should also be within a limited range. The use of chromite in refractory industries has decreased considerably over the last few decades mainly due to its hazardous nature, the changes in steel-making technology, and availability of more suitable refractories. However, it has still an important niche in the refractories industry for certain specific applications.

10.2 RAW MATERIALS AND SOURCES

This chapter details about three different types of chromium-containing refractories, namely, chromite, chrome–mag, and mag–chrome refractories. The main raw

DOI: 10.1201/9781003227854-10

materials used for making these three types of refractories are magnesia and chromite. We have already covered raw material sources of magnesia (in Chapter 8.2), so only chromite raw materials are described here.

In nature, chromium is available mostly as oxide, and is always mixed with various other oxides and silicates. Chromium (Cr) is the 13th most common element in the earth's crust with an average concentration of about 185 ppm. At standard temperature and pressure, chromium is a metal; it does not, however, occur in the native state. The most important chromium-bearing mineral is chromite, which is the only commercially viable ore of chromium. It has a spinel structure with the general formula of AB_2O_4, where A is a divalent (Fe for chromite) and B is a trivalent (Cr for chromite) metal ion.

Chromite ore predominantly contains the iron chromate spinel but is usually associated with the considerable amount of gangue materials, most commonly magnesium silicates. The naturally occurring chromite is a chromium-bearing mineral with a generalized formula of (Mg, Fe)O. (Cr, Al, Fe)$_2O_3$. Magnesium ion replaces the divalent iron, and aluminum and ferric ions substitute the chromium ions. These substitutions can also be seen as mixing of different spinels, namely, iron chromate ($FeCr_2O_4$), magnesium chromate ($MgCr_2O_4$), magnesium aluminate ($MgAl_2O_4$), iron aluminate ($FeCr_2O_4$), etc. Other than these, the mineral may be associated with clay and silicates, which also increase the silica content in the mineral.

Chromite commonly occurs in mafic and ultramafic rocks, which cover large portions of the earth's surface. Peridotite, an ultramafic rock predominated by olivine (magnesium iron silicate), is the common host for economic chromite mineral deposits. Peridotite contains on an average 1%– 2% chromite mineral as an accessory one. Most economic chromite mineral deposits contain concentrations of at least 25% chromite mineral.

Globally, chromite is available in a distributed manner, and the southern Ural area of Kazakhstan is the largest depositary of chromite. As per country wise availability, the important resources for chromite are Kazakhstan, Zimbabwe, South Africa, Finland, India, and Brazil. About 15% of the total world chromite consumption is from the refractory industry. A typical analysis of a chromite suitable for the refractory purpose is 40%–55% Cr_2O_3, 12%–24% Al_2O_3, 14%–24% Fe_2O_3, 14%–18% MgO, and less than 10% SiO_2. As per the chemical constituents, refractory-grade chromite ore is graded into three categories, as detailed in Table 10.1. Grade 1 will result in the best quality of refractories.

TABLE 10.1

Chemical Constituents (% by mass) of Different Chromite Ores Used in Refractory Industries

	Grade 1% by Mass	Grade 2% by Mass	Grade 3% by Mass
Cr_2O_3	52 min	50 min	48 min
Total iron (as FeO)	16 max	16 max	16 max
MgO	15 max	15 max	15 max
SiO_2	3 max	7 max	9 max
Loss on ignition	1.5 max	1.5 max	1.5 max

10.3 BRIEF OF MANUFACTURING TECHNIQUES

10.3.1 CHROMITE REFRACTORY

Hard and lumpy chromite ore is first crushed and ground to the desired particle fractions and beneficiated by physical separation for removal of dust and clayey materials. Commonly, tar is added as a bonding material to bind the ore particles together. Little heating during mixing of tar is required for proper flowability of the mix (due to viscous nature of tar) and better homogeneity. Water-based binders may also be used. Mixed composition is then pressed into shapes and dried slowly to avoid any crack generation. Firing can be done in batch or continuous type of kilns, but a slow firing is essential. Firing temperature varies between 1400°C and 1500°C, depending on the impurities present.

Precautions are required during firing. Chemical reactions occur during firing due to the formation of different types of spinels among the oxide constituents present in it. Mainly, the spinel formation (iron chromate) is associated with an expansion, which causes the expansion of the shape and bursting or shattering. This problem is termed as "cauliflower expansion" (under the microscope the expansion appears to be like cauliflower). To accommodate the expansion in the stack of the refractories, the chromite bricks are stacked loosely inside the furnace with enough gap. Also, to control and accommodate the expansion slowly within the refractory, heating is done slowly.

The addition of magnesia is done to control the cauliflower expansion and also for the formation of magnesium-chromate spinel. Again, magnesia helps to increase the low refractoriness of the gangue material (impurities) by forming magnesium-bearing compounds and improves the hot properties of the chromite refractories.

10.3.2 CHROME–MAGNESIA REFRACTORIES

The fraction of magnesia and chromite ore are mixed to reach the desired composition of the refractory in an edge runner mill or a pug mill with a green binder. Mixed compositions are pressed in hydraulic press at 80–100 MPa pressures and then dried below 80°C to avoid any hydration cracks. Dried shapes are then fired in batch or continuous kilns between 1500°C and 1700°C, depending on the compositions and the impurities present. Heating and cooling are done slowly to avoid any cracking. Here also, the precautions taken for pure chromite refractory is followed to accommodate any expansion that occurs during firing. There are also some unfired chemically bonded refractories that develop strength due to bonding materials present after drying and curing.

10.3.3 MAGNESIA–CHROME REFRACTORIES

Magnesia and chromite are processed separately for crushing and grinding to get different fractions, as required for the desired compaction, and then mixed in desired proportions to obtain the proper composition. Separate minerals/chemicals may be added during the mixing process to form a chemical bond within the refractory.

Mixing is done with water and then dried slowly between 60°C and 80°C to avoid any hydration and hydration-related crack generation. Dried products are fired in batch or continuous types of kilns, and firing are done up to the maximum temperature of 1500°C–1700°C. To accommodate any expansion from spinel formation in the composition, firing is done in a controlled manner at a slow rate. Chemically bonded refractories are not fired but cured for strength development.

10.4 CLASSIFICATIONS AND PROPERTIES

10.4.1 CHROMITE REFRACTORIES

Demand and application of pure chromite refractories are very limited, and so there are nearly no subclasses of chromite refractories. The properties of chromite refractories depend on the amount and types of impurities present, and also on the control of the bursting character (due to expansion) during firing. In the chromite refractories, on firing, FeO present will oxidize to form Fe_2O_3. In reality, this causes an imbalance in divalent and trivalent cation ratio of original spinel structure of the chromite. And, the total amount of spinel will change. In the final fired products, two distinct crystalline phases would appear: one is the spinel phase with MgO (and remaining FeO) and trivalent oxides (like Cr_2O_3, Fe_2O_3, etc.), and the second one is the solid solution of different trivalent oxides (like Cr_2O_3, Fe_2O_3, Al_2O_3, etc.).

As chromia has a high specific gravity, the bulk density of these refractories are high, generally varies between 3.8 and 4.0 g/cc, and apparent porosity is between 24% and 28%. The refractoriness depends on the impurities present and is generally above 1900°C; refractoriness under load varies between 1500°C and 1550°C. The presence of free silica in chromite is highly detrimental to its thermal shock-resistance properties. Chromia is neutral in character. Hence, less-impure chromite refractories show good resistance to chemical attack in different environments. They are resistant to basic slags (except reducing conditions) and fluxes. But the acidic slag and blast furnace slags slowly corrode the refractory.

10.4.2 MgO–Cr₂O₃ REFRACTORIES (CHROME–MAG AND MAG–CHROME)

The major advancement in the technology of basic refractories started after about 1930s, when fired composition of a combination of dead-burnt magnesite and chromite showed improved character than that of the individual components. When a mixture of magnesia and chrome is fired, then the reduction of divalent cation (due to oxidation of iron) is balanced by free MgO present. MgO forms spinel structure with the oxides of trivalent cations present in chromite and also with the newly formed Fe_2O_3. This MgO-containing spinels are high-melting compounds and increase the application temperature of the composition with improved hot properties. The spinels that may form in the $MgO–Cr_2O_3$ refractory compositions are $FeCr_2O_4$, $MgCr_2O_4$, $MgFe_2O_4$, $FeAl_2O_4$ and $MgAl_2O_4$. All these spinel phases are stable up to their high-temperature melting points. They are also congruent in character, which means that they do not decompose and melt to the liquid of its composition. Phase diagrams related to these spinel compounds are given in Figure 10.1 Also, silica is a common

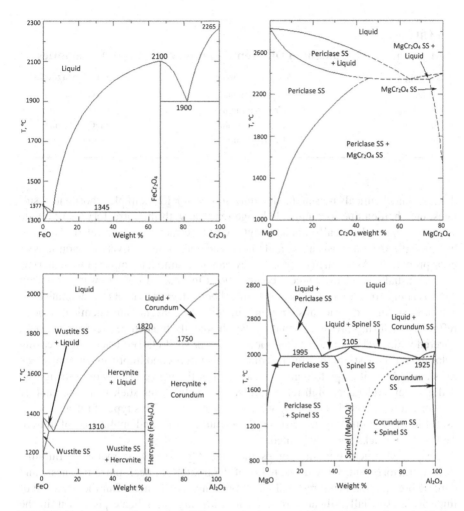

FIGURE 10.1 Phase diagram of different spinels compositions that may form in MgO–Cr₂O₃ refractories.

impurity present in chromite ore, causing low-melting silicate compounds. But the addition of free MgO also converts it to high-melting forsterite (2MgO SiO₂) phase, and the compositions have higher hot properties.

In MgO–Cr₂O₃ refractories, both in chrome–mag and mag–chrome types, the classification is done as per the bond present in the refractory. The main bonding used in these refractories are sintering bond or chemical bond. The chemical compositions and properties of MgO–Cr₂O₃ refractories also change with the bonding system, and the details of the amount of MgO and Cr₂O₃ content with the bonding system are given in Table 10.2.

The simplest way to develop a bonding is the chemical bonding, which is basically unfired bricks. In chemical-bonded refractories, pressed shapes are dried, and

TABLE 10.2

MgO and Cr_2O_3 Content in Different Types of MgO–Cr_2O_3 Refractories

Bonding	Magnesia Chrome	Chrome Magnesia
Chemical	MgO–55% min	MgO–30% min
	Cr_2O_3–6% min	Cr_2O_3–18% min
Sintering	MgO–75% min	MgO–50% min
	Cr_2O_3–6% min	Cr_2O_3–18% min

the bonding chemicals present (added during mixing), for example, phosphates, sulfates, etc., harden due to chemical change/reaction with the refractory and provide strength. These chemically bonded MgO–Cr_2O_3 refractories have relatively lower hot strength and slag resistance, and are considered to be relatively economic with poor properties. As no firing is done, they can accommodate impure raw materials and are usually used in low-cost compositions to balance out the wear profile in various application areas. But as the chemical bond degrades at an intermediate temperature, strength deteriorates. In a modification to conventional chemical-bonded refractories, during pressing, some of the chemically bonded refractories surfaces covered with steel plates are called metal-cased or steel-classed brick. Metal casing helps to retain the strength at intermediate temperatures and improves thermal shock resistance. Also, it helps to maintain the bond in the whole brick work during use, and as steel oxidizes at high temperatures, the metal cases stick with one another and form a tight bond between the whole brick structure. This type of refractories is useful for roof applications. However, the manufacturing and application of unfired MgO–Cr_2O_3 bricks are very limited.

Next to chemical bonding comes the fired MgO–Cr_2O_3 refractories. Initially, these refractories started with the firing of the mixture of dead-burnt magnesia and chrome ore, and the firing resulted in the bonding and strength. But the presence of impurities, especially silicates, resulted in a low-melting silicate phase within the refractory. A thin film of silica-based coating on magnesia and chrome particles provide the strength. The firing of these refractories is dependent on the amount and types of impurities and silicate phases present in the system. Again, due to the formation of low-melting liquids at high temperatures from impurities and silicates, these refractories have low hot strength and corrosion resistances. Also, the presence of silicate-based glassy bond phase results in a relatively lower thermal shock resistances. Though the development of fired MgO–Cr_2O_3 started with silicate bonding, it changed slowly to direct-bonded refractories and is presently used in very limited areas.

Use of very high-purity raw materials or reduction in impurity content from the starting magnesia and chromite will minimize the silicate phase formation between the magnesia and chromite grains in the fired refractories. Hence, these magnesia and chromite grains will bond directly with each other to form the direct bonding. These directly bonded refractories are only possible by using high-purity raw materials and firing the pressed bodies to very high temperatures (>1700°C). The higher the temperature and longer the firing schedule, the greater extent of reaction will take place, and greater bonding will occur with a higher amount of spinel formation. Both the primary and secondary spinels will be found in the microstructure of the product, and the refractory will be volume stable, and has high hot strength with great corrosion and thermal shock resistances. But obviously, these refractories are relatively costly compared to that of the chemically bonded or silicate-bonded ones.

There is also another variety of $MgO-Cr_2O_3$ refractory termed as co-burnt or rebonded refractory. In this case, high-purity raw materials (in fine form) are mixed as per proportions required to obtain desired composition and compactness, pressed to pellet, and co-burnt (dead burning) in kilns (say rotary kiln) at very high temperatures. The resulting grains are dense and directly bonded in nature. These grains are used for making the shaped refractory by mixing in different fractions, pressing, and firing. As the expansion due to spinel formation has already occurred during co-burning, these refractories shrink on firing and do not expand like the previous ones. These refractories are high volume stable, as nearly all the reactions (spinel formation) are completed in two different high-temperature firings, has the high hot strength and other hot properties, and has much lower porosity, resulting in further improvement in corrosion resistances.

In a special variety of $MgO-Cr_2O_3$ refractories, fused magnesia–chrome grains are used especially to improve the corrosion resistance. Dead-burnt magnesia and chrome ore are first mixed as per desired proportions; then the mix is fused in an electric arc furnace, cooled, and then crushed and ground to get desired fractions of fused grains for making refractory shapes. The shapes are then fired at high temperatures. As the reactions within the batch materials have already been completed during fusion, these shapes also do not expand but shrink during the refractory firing. Due to the presence of fused grains, these refractories have lower porosity, very high corrosion resistances, and hot strength properties.

In general, properties of the $MgO-Cr_2O_3$ refractories depend on the composition, impurities present, and types of the bond phase present. Due to the presence of Cr_2O_3 in the compositions, all the refractories look gray to dark brown in color, depending on the amount of Cr_2O_3 and Fe_2O_3 present.

Refractoriness of these refractories are generally above 2000°C, and the RUL value is above 1700°C for the direct-bonded, rebonded, and fused-grained bricks. But due to the presence of the impurity phases and formation of the liquid phase at lower temperatures in chemically bonded and silicate-bonded refractories, the RUL value is relatively low and varies between 1500°C and 1550°C.

Thermal spalling properties of $MgO-Cr_2O_3$ varies with the amount of MgO and Cr_2O_3 content and the impurities present in the composition. Chrome–mag refractories have better spalling resistance than the individual magnesia or chrome refractories, but have a relatively lower thermal spalling resistance than magnesia–chrome refractories.

Regarding corrosion, refractories with chemical bonding and silicate bonding have poor resistance compared to that of the direct bonded, fused grains and rebonded $MgO-Cr_2O_3$ refractories. The presence of impurities and formation of the liquid phase within the refractories strongly deteriorate the corrosion properties. Again, when the compositions have higher Cr_2O_3 content, for example, in chrome–mag compositions, the refractories disintegrate in the presence of higher FeO in the slag. This may be the due absorption of FeO and further formation of iron chromite, which causes expansion of the refractory. The more is the amount of FeO, the more is the absorption, and the higher is the degree of deterioration. Also, the presence of CaO in the system (e.g., in slag) further deteriorates the situation. Compared to chrome–mag refractories, mag–chrome refractories show better performances against FeO corrosion. But at a very high concentration of FeO in slag or very high content of iron in chromite itself causes similar deterioration in the mag–chrome refractories. Hence, among the different compositions of $MgO–Cr_2O_3$, high-magnesia-containing (e.g., 80%) refractory is strong against basic slag attack and alkali environment, whereas a composition with a lesser amount of MgO (55%–60%) and high Cr_2O_3 (18%–25%) is stronger in the thermal spalling environment.

10.5 MAIN APPLICATION AREAS

Pure chromite refractories are available and used in a very limited manner. They are mostly used in reheating furnaces, like hearth of soaking pits and in rolling mills for their high abrasion resistance. High Cr_2O_3 (95%)-containing dense refractories are important for fiber glass making furnaces in glass contact areas due to their excellent corrosion resistance. However, there are issues related to coloring problem.

Use of $MgO–Cr_2O_3$ refractories has started around 1930, and the development continued till the end of the last century. They were most important for the basic open hearth furnaces and electric arc furnaces. But the application of these refractories has decreased greatly with the development and increased use of MgO–C refractories, reduction in steel making through open hearth furnaces, and introduction of water cooling technology for walls and roofs of electric arc furnaces. Still $MgO–Cr_2O_3$ refractories are important in many other applications, which are described as below.

Magnesia–chrome refractories are commonly used in secondary steelmaking process due to the resistance against various basic slags and volume stability at high temperatures. Relatively low-grade mag–chromes are used as back up lining in steel converter, steel ladle, etc. Direct bonded and co-burnt mag–chrome refractories are important for the degassing furnaces, like AOD and VOD, electric arc furnaces, etc. As on today, degassers are the largest consumer of magnesia–chrome refractories. Co-burnt mag–chrome is useful for steel vacuum degasser. Chrome–magnesite refractories are used for the sidewalls of the electric arc furnaces, soaking pits, and for the lining of induction furnaces. Electric arc furnaces (EAFs) used by the mini steel plants are conventionally lined with fused-cast mag–chrome refractories in the high-wear hot spots (walls nearest to the electrodes), direct-bonded and co-burnt mag–chrome in the remainder of the sidewalls, and slag lines and chemical-bonded and direct-bonded magnesia–chrome brick in the roofs. EAFs used in stainless steel production also uses direct-bonded and rebonded fused-grain magnesia–chrome (60 wt% MgO) brick in the slag line and lower hot-spot areas of sidewalls.

Different furnaces used for making copper use $MgO–Cr_2O_3$ refractories. The linings of copper converter and copper reverberatory furnaces are done with chrome–mag refractories. Again, fused-cast $MgO–Cr_2O_3$ composition is used in converter furnace bottom and tuyere zone, having the highest wear rate, to minimize the corrosion. Tuyere area of copper converter has a higher rate of wear and is generally lined with a better-quality refractory compared to that of the rest of the converter. Like, if silicate-bonded chrome–mag is used for the bulk lining, then direct-bonded mag–chrome is to be applied for the tuyere, but if direct-bonded refractory is used for the bulk lining, then rebonded or fused-grain mag–chrome is to be used for tuyere areas.

Direct-bonded high-temperature fired-high MgO(~80%–85%)-containing mag–chrome refractories with little iron oxide are used in the burning and transition zones of cement rotary kilns due to their excellent strength, high volume stability, corrosion resistance, and easy coating formation.

Co-burnt high-fired mag–chrome refractories are important for various nonferrous metal industries like lead, zinc, and nickel furnaces due to their excellent corrosion, erosion, and thermal shock resistances.

10.6 HAZARDS WITH CHROMITE-CONTAINING REFRACTORY

It has been found by many research works that chromite-containing refractories form toxic compounds when they are exposed to high temperature, high pressure, chemical contacts, etc. Hexavalent chromium compounds, exceeding the EPA (United States Environment Protection Agency) limits, were found in chrome-bearing materials when they are in contact with calcium aluminates.

Environmentally hazardous, hexavalent, chromium-containing CrO_3 forms in refractories along the grain boundaries. The transition of Cr^{3+} into Cr^{6+} in the air is accelerated when chromite is in contact with alkali and alkaline earth oxides. Hence, chromite-containing refractories must be avoided in contact with alkali, alkaline earth oxides, especially calcium oxide. Formation of Cr^{6+} formation may also be controlled by controlling the temperature, basicity (CaO/SiO_2), and chromite particle size used in making the refractory. The use of coarser chromite particles and fused grains resulted in decreased formation of Cr^{6+} during refractory use.

Again most of the refractories are disposed as landfill after use. Any conversion to Cr^{6+} will leach out in such disposal and is a great threat to the environment. Recycling, reuse, and waste management of refractories are extremely difficult, since these processes are associated with cost, difficulty in separation of contaminated part, quantify the quality product, etc. Also, due to the risk of contamination, recyclable, spent refractories are low-value items and rarely get proper attention. A great thrust is required to recycle and reuse these chromite refractories, not only to protect the environment but also to make the process economic. Also, plenty of work has been done to substitute chromite-containing refractories in almost all the application areas and in many advanced countries who have banned the use of Cr_2O_3-containing refractories as a safety measure.

SUMMARY OF THE CHAPTER

Chromite-containing refractories are mainly of three types, namely, chromite, magnesia–chrome, and chrome–magnesia refractories.

Chromite is a natural mineral that ideally contains FeO and Cr_2O_3. But, in natural occurrences, Fe is replaced mainly by Mg, and Cr is replaced mainly by trivalent Fe and Al ions. Also, it is contaminated with silica.

$MgO–Cr_2O_3$ refractories use magnesia mainly in the pure dead-burnt form and chromite as a natural source of raw material.

Chromite refractories have nearly no classification. But, for $MgO–Cr_2O_3$ refractories, as per the content of Cr_2O_3, they are classified as mag–chrome and chrome–mag.

$MgO–Cr_2O_3$ refractories are also classified as per the bonding, namely, chemically bonded, silicate bonded, direct bonded, co-burnt/rebonded, and fused grains types.

Chromite refractories, and chemically bonded and silicate-bonded $MgO–Cr_2O_3$ refractories have higher impurity content, and result in relatively poor hot properties and corrosion resistances.

$MgO–Cr_2O_3$ refractories are mostly used ferrous and nonferrous metallurgical industries, cement kilns, and glass industries.

At high temperature, trivalent chromium may convert to hexavalent state, which is a toxic material, and the presence of alkalis and calcium enhances the conversion. Hence, use and disposal of chromium-containing refractories need special attention.

QUESTIONS AND ASSIGNMENTS

1. Discuss in detail about the chromite ore.
2. Detail about the manufacturing, properties, and applications of chromite refractories.
3. Discuss the different classifications of $MgO–Cr_2O_3$ refractories.
4. What are the different bonding systems used in $MgO–Cr_2O_3$ refractories? Discuss them.
5. Discuss the advantages and disadvantages of different bonding system in $MgO–Cr_2O_3$ refractories.
6. Discuss the difference in manufacturing techniques of different $MgO–Cr_2O_3$ refractories having different bonding.
7. Mention the applications of chromite refractories.
8. Discuss the applications $MgO–Cr_2O_3$ refractories.
9. What are the dangers of chromium-bearing refractories? Discuss the same.

BIBLIOGRAPHY

1. J. H. Chesters, *Refractories- Production and Properties*, Woodhead Publishing Ltd, Cambridge, 2006.
2. P. P. Budnikov, *The Technology of Ceramics and Refractories*, Translated by Scripta Technica, Edward Arnold, The MIT Press, Massachusetts, US, 4th Ed, 2003.
3. *Harbison-Walker Handbook of Refractory Practice*, Harbison-Walker, Moon Township, PA, 2005.
4. C. A. Schacht, *Refractories Handbook*, CRC Press, Massachusetts, US, 2004.
5. A. Rashid Chesti, *Refractories: Manufacture, Properties and Applications*, Prentice-Hall of India, New Delhi, 1986.
6. W. David Kingery, H. K. Bowen and Donald R. Uhlmann, *Introduction to Ceramics*, John Wiley and Sons Inc., New York, 2nd Ed., 1976.

7. Waing Waing Kay Khine Oo, Shwe Wut Hmon Aye and Kay Thi Lwin, Study on the production of chromite refractory brick from local chromite ore. *World Academy of Science, Engineering and Technology*, 46 569–574 (2008).

8. Waing Waing Kay Khine Oo, Shwe Wut Hmon Aye and Kay Thi Lwin, *International Journal of Chemical, Molecular, Nuclear, Materials and Metallurgical Engineering*, 2 [10] 234–239 (2008).

9. C. Kim, A. Kohler, K. Mulvaney and L. Wagner, *Ceramics Industry*, 142 [8], 57–61 (1992).

10. *Technology of Monolithic Refractories (Plibrico Company)*, p. 69, Toppan Printing Company, Tokyo, Japan, Japan, 1985.

11. D. J. Bray, Toxicity of chromium compounds formed in refractories. *American Ceramics Society Bulletin*, 64 [7] 1012–1016 (1985).

12. C. G. Marvin, Chrome bearing hazardous waste. *American Ceramics Society Bulletin*, 72 [6] 66–68 (1993).

13. L. B. Khoroshavin, V. A. Deryabin, V. A. Perepelitsyn and T. N. Lapteva, Hexavalent chromium in refractories. *Ogneupory*, 58 [9] 7–10 (1993).

14. J. P. Bennett and M. A. Maginnis, Recycling/disposal issues of refractories. *Ceramic Engineering and Science Proceedings*, 16 [1], 127–141 (1995).

15. B. F. Belov, I. A. Novokhatskii, L. N. Rusakov, A. V. Gorokh and A. A. Savinskaya, Phase diagram for an iron (II) oxide-chromium (III) oxide system. *Russian Journal of Physical Chemistry A*, 42 [7] 856–858 (1968).

16. A. M. Alper, R. N. McNally, R. C. Doman and F. G. Keihn, Phase equilibria in the system $MgO-MgCr_2O_4$. *Journal of the American Ceramic Society*, 47 [1] 30–33 (1964).

17. I. A. Novokhatskii, B. F. Belov, A. V. Gorokh and A. A. Savinskaya, The phase diagram for the system ferrous oxide-alumina. *Russian Journal of Physical Chemistry A*, 39 [11] 1498–1499 (1965).

11 Magnesia–Carbon Refractories

11.1 INTRODUCTION

Magnesia–carbon refractory is a composite type of unfired refractory essentially required for the modern steel-making process. As the refractory contains carbon (graphite as a source), they are not fired, otherwise, carbon will oxidize at the ambient firing conditions. Also carbon/graphite, being a hydrophobic material, it does not disperse in water-containing system, so for proper and uniform distribution of carbon with magnesia organic binders are used instead of water, like pitch, tar, resin, etc. These binders polymerize under certain conditions, retain the shape, and provide strength. As, on today, MgO–C refractories are essential for both the primary and secondary steel-making processes.

The initial development and first use of basic refractories mixed with carbon were started in the early 1950s. These early carbon-containing basic refractories resulted in only about 100 heats in converter. Poor repair technology and huge production pressure were the reasons for reduced service life. Considerable improvement in the performance and life were obtained in the early days, when magnesia fines were used in combination with the dolomite coarse fractions bonded with pitch. Further improvements were obtained when dolomite was replaced by magnesia in pitch-bonded compositions. Later, in the 1970s, pitch-impregnated burnt magnesia bricks having reduced surface pore and pore size were developed for the charge pad and other high-wear areas. This was the initiation of the zonal lining concept in basic oxygen furnaces. Slowly, purity of magnesia became an important factor, and high, pure (>96%), low, boron-containing magnesia with lime–silica ratio above 2 became the preferred material.

After 1980s, slowly the conventional binders like pitch and tar were replaced by resin mainly due to higher carbon (fixed) content and environmental friendliness. Use of additives like, antioxidants, also started to improve the performance and service life. High-quality magnesia and carbon sources satisfying certain specific criteria were being used to improve further the corrosion resistance and other hot properties.

At present, the MgO–C refractories are prepared by using high-quality magnesia and carbon sources, and bonded by high carbon containing an organic binder (mainly resin), with some metallic powder as antioxidants to protect the carbon. These MgO–C refractories are shaped at high pressure and are tempered/cured but not fired. The main features of MgO–C refractories are:

1. High refractoriness as no low-melting eutectic occurs between MgO and C.
2. Graphite, the carbon source, has very low thermal expansion; hence, the composite has also a low expansion value.

3. Graphite, having an unshared free electron, has very high thermal conductivity, which imparts high thermal conductivity in the MgO–C composite.
4. As the thermal expansion is low, and the thermal conductivity is high, the thermal shock resistance of MgO–C refractories is very high.
5. As carbon molecules have very strong covalent bonding, they do not react much and do not get wet by any liquid. This nonwetting character results in very high corrosion resistance in MgO–C refractories against molten metal and slag.
6. Graphite has a low modulus of elasticity, hence ability to absorb stress with minimum deformation, thus keeping down the amount of discontinuous wear due to cracks.

Conceptually, the presence of carbon in magnesia refractory can be termed as magnesia–carbon (MgO–C) refractory. So, under the broad description of carbon-containing magnesia refractories, there are three subclasses:

- sintered and impregnated (pitch/tar) magnesia, containing up to 3% carbon,
- tar/pitch-bonded magnesia, containing up to 7% carbon, and
- carbon (graphite)-containing magnesia, with carbon as a constituent batch material present in the range of 8–30 wt%.

But conventionally, the third category of refractories are considered as MgO–C refractories and being considered all over the world.

11.2   RAW MATERIALS, BINDERS, AND ADDITIVES

Magnesia–carbon refractories are made up of two major components, namely magnesia and carbon. Other than these, antioxidants are used as additives, and an organic binder is required for shaping and providing strength after curing/tempering.

11.2.1   MAGNESIA

Magnesia constitute the major portion of magnesia–carbon refractory, varying between 80% and 90% by weight. Three different types of magnesia are used for magnesia–carbon refractories, namely, fused magnesia, produced by fusing magnesia in an electric furnace; seawater magnesia, produced by very high-temperature firing of magnesium hydroxide extracted from seawater; and sintered magnesia produced from natural magnesite. Details of the magnesia raw materials are discussed in Section 8.2. Being the major part, magnesia plays the pivotal role in the property development and performance of the MgO–C refractory. There are different selection criteria for magnesia to improve the quality of the refractory. These criteria are as follows:

a. High purity
b. High ratio of CaO/SiO_2
c. Minimum content of B_2O_3
d. Large periclase crystal size to reduce the grain boundary area.

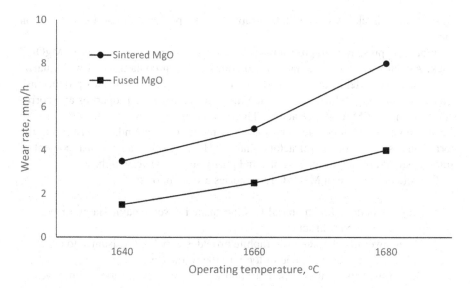

FIGURE 11.1   Effect of fused and sintered magnesia grains on the wear rate (corrosion) of the MgO–C refractory.

Use of impure magnesia will produce low-melting phases in the system, degrading the hot strength and corrosion properties against molten metal and slag. The similar effect will also be there if magnesia with lower lime–silica ratio or higher B_2O_3 content is taken. Use of fused grains results in increased periclase crystal sizes and provides higher corrosion resistances. Figure 11.1 shows the effect of fused and sintered magnesia grains on the wear rate (corrosion) of the MgO–C refractory. Increased temperature enhances the rate of corrosion due to higher slag fluidity. Fused magnesia-containing refractory showed much higher corrosion resistance than the sintered ones. But larger-sized crystals may affect the mechanical and thermal shock resistance properties due to reduced number of grain boundaries and less resistance to crack propagation. So a mixture of fused and sintered magnesia is generally used, and the proportion is optimized as per the specific application area requirement.

11.2.2   Graphite

Carbon plays very vital role in MgO–C refractories due to its nonwetting character. But carbon has the instantaneous oxidation tendency at high temperatures in oxidizing atmosphere, generally prevailed in the user industries. Oxidized MgO–C refractory has a porous structure, with very low strength, and corroding liquids can enter in it, resulting in a drastic deterioration of the refractory quality. Among different commercial sources of carbon, graphite shows the highest oxidation resistance. So it is selected as the source of carbon in MgO–C refractories. Again, as the flaky nature imparts higher thermal conductivity and lower thermal expansion, resulting in very high thermal shock resistance, flaky graphite is the preferable material as a carbon source.

Impurities present in graphite adversely affect the corrosion resistance of MgO–C brick. As flaky graphite is a natural mineral, it is often contaminated with quartz, muscovite, pyrite, feldspar, iron oxide, kaolinite, etc. These impurities produce ash after oxidation of graphite, and degrade the high-temperature properties and corrosion resistance of MgO–C refractory. These impurities may react with MgO or with the slag component in contact, and form low-melting compounds, deteriorating the corrosion resistance of the refractory sharply. Hence, impurities and ash content of graphite should be as low as possible, and purity should be very high.

The role of graphite in MgO–C refractories are as follows:

1. Graphite being a fine material fills the space between magnesia grains and produces compact structure.
2. The nonwetting character of graphite provides very high resistance to metal and slag corrosion and penetration into the refractory.
3. Due to high thermal conductivity and low thermal expansion, it improves the thermal spalling resistance.

11.2.3 ANTIOXIDANT

Oxidation of carbon is the main drawback of carbon-containing refractories. The oxidation of carbon in MgO–C refractories may occur in two ways, namely, (i) direct oxidation and (ii) indirect oxidation. Direct oxidation occurs below 1400°C, when carbon is oxidized directly by the oxygen from the atmosphere.

$$2C \; (S) + O_2 \; (g) = 2CO \; (g)$$

Indirect oxidation occurs generally above 1400°C, where carbon is oxidized by the oxygen from MgO or slag (mainly FeO). The resulting Mg gas moves through the oxidized porous refractory structure and comes out to the open environment from the refractory surface. At the refractory surface, the Mg gas comes in contact with air (oxygen), gets oxidized, and forms MgO (secondary oxide phase). This nascent MgO gets deposited on the hot face of the refractory, forms dense layer that acts as an oxide coating, and prevents the penetration of oxygen within the refractory and so prevents further oxidation.

$$C \; (S) + MgO \; (S) = Mg \; (g) + CO \; (g)$$

$$C \; (S) + FeO \; (l) = Fe \; (l) + CO \; (g)$$

$$2Mg \; (g) + O_2 \; (g) = 2MgO \; (s)$$

But, in both the cases, carbon gets oxidized, and refractory becomes porous, weak in strength, and gets corroded easily. To prevent the oxidation of carbon, antioxidants are used. Antioxidants are materials that get oxidized faster by reacting with an incoming oxygen and protects the carbon from oxidation. These materials are simple metal powders that oxidize easily and faster. Thus, they consume the incoming oxygen and protect the carbon from oxidation. Generally, powders of magnesium (Mg), aluminum (Al), silicon (Si) metals, or fine boron carbide (B_4C), silicon carbide (SiC), etc. are used as antioxidants. Due to the low cost and high effective protection, Al and Si metal powders are mostly used, which after oxidation form their respective oxides, and remain as a stable discrete phase in the refractory, even at high temperatures. Also, the formed metal oxides have greater volume and cover the carbon particles to provide greater oxidation resistance even at high temperatures.

$$4Al\ (S) + 3O_2\ (g) = 2Al_2O_3\ (s)$$

$$2Mg\ (S) + O_2\ (g) = 2MgO\ (s)$$

$$Si\ (S) + O_2\ (g) = SiO_2\ (s)$$

During operation at high temperatures, carbon also reacts with the metal powders, forming metallic carbides. For example, in Al-containing compositions, Al_4C_3 forms. This reaction provides extra bonding to the refractory, resulting in further improvement of strength. Al metal powder melts around 660°C, which then reacts with carbon and forms the carbide.

$$4Al\ (l) + 3C\ (s) = Al_4C_3(s)$$

This carbide formation is especially effective in increasing the hot strength of the MgO–C refractory. Figure 11.2 shows that metal additives enhance the hot strength

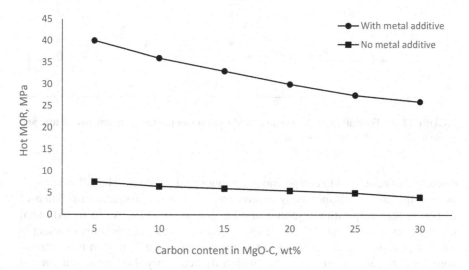

FIGURE 11.2 Variation of hot strength values of MgO–C refractory with and without metal additives against carbon content.

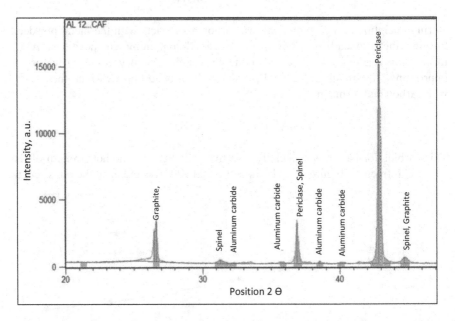

FIGURE 11.3 Formation of aluminum carbide and spinel phases in the microstructure and phase analysis of MgO–C refractories.

considerably compared to compositions without metal additives. The fall in strength in both the cases with increasing carbon content is due to enhanced oxidation of carbon in the composition resulting in more porous structure. Again for Al metal powder-containing compositions, the formed Al_2O_3 reacts with MgO fines present in the matrix phase and forms $MgO.Al_2O_3$ spinel. This spinel formation is associated with a volume expansion that occurs within the refractory. This increased volume fills up of the pores and reduces porosity. This reduced pore volume helps in reduction of the oxidation and corrosion of the refractory.

$$4Al\ (s) + 3O_2\ (g) = 2Al_2O_3\ (s)$$

$$Al_4C_3\ (s) + 6CO\ (g) = 2Al_2O_3\ (s) + 9C(s)$$

$$Al_2O_3(s) + MgO\ (s) = MgO.Al_2O_3(s)$$

The formation of aluminum carbide and spinel phases are also observed in the matrix part of MgO–C refractories, as shown in Figure 11.3. Phase analysis also indicates and confirms the formation of these phases, but the intensity of these phases are relatively low due to the very little extent of the formation in the matrix phase.

As an antioxidant Al, metal powder is preferred and widely used, given the excellent high-temperature strength of the refractory and its relatively low cost. However, MgO–C refractory with Al metal powder as antioxidant suffers from some shortcomings too. At low temperatures, formed Al_4C_3 reacts with the water vapor and expands enormously, forming crack and reducing the service life. Formation of voluminous $Al(OH)_3$ and CH_4(methane) gas causes the expansion and disintegration of the refractory.

$$Al_4C_3(s) + 12H_2O(g) = 3CH_4(g) + 4Al(OH)_3(s)$$

This is especially important for refractory linings when they are cooled to ambient temperature for intermittent repairs, which may take several days. In rainy seasons, humidity is high, and this may cause serious damage or deterioration of the whole refractory lining due to this hydration.

Boron carbide behaves in a different manner as an antioxidant. It reacts with oxygen or CO, and forms B_2O_3. This B_2O_3 is a low-melting material, and as a liquid it coats the carbon particles, and thus prevents the oxidation. Also, it reacts with MgO and forms low-melting liquid phase compound $MgO.B_2O_3$.

$$2B_4C(s) + 6O_2\ (g) = 4B_2O_3(l) + 2C\ (s)$$

$$B_4C(s) + 6CO(g) = B_2O_3(l) + 7C\ (s)$$

$$B_2O_3(l) + 3MgO(s) = Mg_3B_2O_6(s)$$

The melting point of $Mg_3B_2O_6$ is 1360°C, and liquid $Mg_3B_2O_6$ fills up the pores. It also acts as a good oxygen barrier above its melting point, as it forms a thin coating on the brick surface that restricts the oxygen diffusion into the refractory. Thus, B_4C protects the carbon of MgO–C bricks in a better way against oxidation, even from a very low temperatures. But the amount of B_4C used has to be judiciously used, and any excess amount will produce greater extent of the liquid phase and will degrade the hot properties of the refractory drastically. The addition of SiC helps in preventing the oxidation of carbon in the similar way as that of B_4C. Silicate liquid phase is formed based on the $MgO–SiO_2$ system that coats the refractory and prevents the oxidation. But an excessive amount of silica will result in excessive liquid phase and is detrimental for the high-temperature properties of MgO–C refractories.

11.2.4 RESIN

It has been already mentioned that water cannot be used for mixing of carbon-containing compositions, so organic liquids are used. Again graphite used has a flaky structure, and so it has very poor compressibility. That is when the pressure is released after shaping or after drying process, graphite will spring back to its original dimensions if it is held tightly and the refractory will expand and crack. Hence, very strong binder is required for MgO–C refractories that can strongly hold the graphite flakes together (as in under pressed condition) after pressing, and retain the pressed shape and dimensions.

Development of magnesia–carbon refractories started with impregnation of liquid pitch and tar in sintered magnesia refractories. This impregnation imparts a coating of carbon on the oxide; refractory surface improves the corrosion resistance. But once the coating of carbon is removed during use (even after little operation or few heats), the oxide refractory is exposed, and no benefit of carbon was present further. These carbonaceous materials, tar or pitch, next were used as an organic binder material in magnesia refractory. But during curing/tempering stage (for carbon–carbon network formation and development of strength) of the tar/pitch-bonded refractories, huge amount of toxic fumes containing polycyclic aromatic hydrocarbons (PAH) such as benzo-alpha-pyrenes (BAP) (carcinogenic material) are liberated. To overcome the crisis, use of phenolic resin has started as bond material replacing tar and pitch. The advantages of phenolic resin are:

- Greater chemical affinity to graphite, so easy dispersion,
- Highly adhesive in nature, resulting in dense shape with good strength,
- A lesser extent of toxic gas evolution,
- Liberates mainly phenol and not PAH compounds during tempering,
- Excellent kneading and pressing properties, and
- Higher amount of fixed carbon.

Phenolic resins are of two types: thermo-setting (resol) and thermoplastic (novolac). Thermo-setting resins give greater dried strength and high resistance against lamination (lamination occurs due to the presence of flaky graphite). Again thermoplastic resins require a chemical to set or harden, called hardener (like hexamine), and they provide greater mix life (conservation of mix quality for a longer time before mixing). Polymerization of the resin occurs due to tempering at ~200°C, and leads to an isotropic interlocking structure that holds the refractory composition in three dimension and provides strength. However, the resins have the limitation regarding their cost in comparison with pitches.

Viscosity (fluidity) of resol resin varies with temperature, and so it behaves differently in different seasons, like winter and summer. Heating may be required during mixing to obtain the proper dispersion of fines (including graphite) in winter, and, again in summer, the green body becomes soft due low viscosity resulting in lamination. To overcome such crisis powder, novolac resin is mixed with the liquid resol resin. The increase in resin content improved compressibility during pressing and corresponding strength of the tempered samples.

11.3 BRIEF OF MANUFACTURING TECHNIQUES

Manufacturing of MgO–C refractories involve no firing/sintering, but the refractory has to get similar kind of compaction and strength, as they are mostly used in the most critical areas of steel manufacturing. Hence, the critical stages of manufacturing these refractories are pressing, where initial compaction is given, and curing/tempering, where the bond material polymerizes to form a three-dimensional carbon chain network providing strength to the refractory even at high temperatures.

In the manufacturing process, first the raw materials, namely, sintered and fused magnesia of different size fractions, graphite, and additives (mainly antioxidant) are mixed, and during mixing organic binder (resin) is added. Mixing continues for about 30 minutes; if required, heating may be done depending on the resin viscosity. Viscosity of resin is maintained in the range of 6000–8000 cps for proper distribution of graphite and other fines. Proper mixer is also required for better mixing performance, as some kneading action is also necessary for better homogeneity in the batch. Generally, small-capacity heavy duty mixers are used to get proper mix quality with a uniform coating of binder on the aggregate particles.

After mixing, the mixture is allowed to age for few hours, which helps in the polymerization of carbon binder and formation of the carbon–carbon interlocking structure. Ageing under controlled atmosphere (maintaining specific temperature, humidity, etc.) is also done in some cases for better and uniform property development within the mix. The aged mixture is then shaped generally by uniaxial pressing in a hydraulic or friction press. The presses should be enabled with the de-airing facility to remove the entrapped air from the batch mixture (to reduce or remove lamination) and capable of very high pressure generation. The durability and performance of MgO–C refractories depend on the compaction, density, and porosity of the body, so a very high pressure, up to about 200 MPa, is used for compaction. After pressing, the shapes are tempered/cured. Tempering is the heat treatment process at low temperature to remove volatile matters from the organic binders and to complete the carbon chain network formation that imparts strength. Tempering is done between 180°C and 250°C, depending on the type and properties of the resin. Flow diagram for the manufacturing process is outlined in Figure 11.4.

11.4 CLASSIFICATIONS AND PROPERTIES

Classification of MgO–C refractories is based on the composition of the refractory, mainly the carbon content. The properties and performance of the refractory change with the amount of carbon and their application areas too. An increasing amount of carbon will reduce the bulk density, as lighter graphite will be replacing denser magnesia in the batch. In parallel, it will also reduce strength, as it hinders any direct bonding between magnesia particles and also elastic modulus, as graphite is a soft material (lower modulus of elasticity). Again, the oxidation of the refractory will increase. But increasing carbon means higher thermal conductivity and lower thermal expansion of the composition that results in greater thermal shock resistance. Also, corrosion and penetration resistances will significantly increase due to the non-wetting character of graphite.

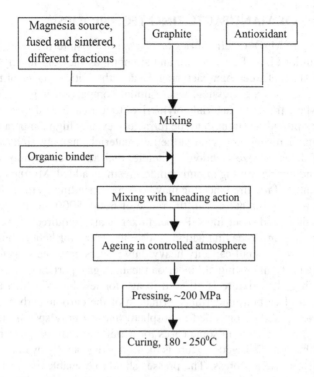

FIGURE 11.4   Flow diagram for manufacturing of MgO–C refractory.

Hence, an optimization in the carbon content is essentially required to get an ideal property for a specific application. It has been observed, as shown in Figure 11.5, that increasing amount of carbon content reduces the corrosion of the refractory, represented as corrosion index, where corrosion of the without-carbon-containing composition is taken as 100. But above 20% of carbon content in the refractory again showed enhancement of corrosion. An increasing amount of carbon enhances the nonwetting character in the composition and reduces the corrosion effectively. But oxidation of the refractory also increases, which produces a porous structure. At the lower amount of carbon, oxidation is less, so effective resistance against corrosion increases with increasing carbon content. But at a higher percent of carbon content, the oxidation is also strong, and the benefit of nonwetting character due to the presence of higher carbon is nullified by oxidation of carbon. Due to oxidation of carbon the refractory becomes porous and easily gets penetrated and corroded. A minimum corrosion is observed for the composition containing about 18–20 wt% of carbon. Any higher amount of carbon causes greater oxidation that results in porous structure and greater corrosion.

Though these refractories are not fired, they are well dense due to high pressing, the bulk density varies between 2.8 and 3.1 g/cc depending on the carbon content, and has low apparent porosity in the range of 4%–6%. Cold strength and hot strength (HMOR) also vary from about 60 and 18 MPa for 8% carbon batch to 30 and 10 MPa for the 20% carbon-containing composition, respectively.

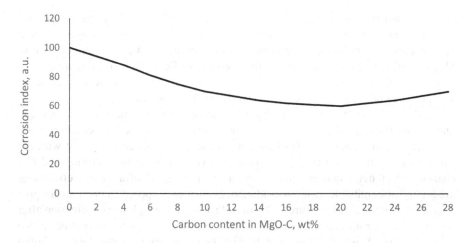

FIGURE 11.5 Effect of carbon content on the performance of MgO–C refractory.

11.5 DEGRADATION OF MGO–C REFRACTORIES

MgO–C refractories are composite in nature and have the benefits of both the components. They are applied in very stringent and critical environments of steel-making process. MgO–C refractories are essential in few applications, and there is no substitute for them with respect to the performance and service life in such critical applications. But still, they degrade due to many reasons. The main reasons for the degradation of MgO–C refractories are detailed below.

11.5.1 DISSOLUTION OF MAGNESIUM BY SLAG

If the basicity of slag is low, then the greater amount of SiO_2 present in slag will react strongly with the basic component (MgO) of the refractory, leading to the formation of magnesium-containing silicates. This dissolution of magnesia in low, basic slags will continue through the grain boundaries, and grain size of magnesia will decrease in contact with silicate slags. Finally, the grain will dislodge from the main refractory. Also, the higher presence of FeO in the slag will also react and dissolve out the MgO from refractory, and slag will penetrate within the periclase crystals. In real situation both the dissolution, mechanism runs parallel and cause degradation of refractory very strongly. The use of purer MgO particles with larger size (fused grains) reduces these problems. Also, controlling the slag chemistry to make the slag highly basic (by increasing the MgO content of slag) restricts such degradation of MgO–C refractory.

11.5.2 OXIDATION OF CARBON

Oxidation of carbon produces a void space in the refractory; the porous structure of refractory is weak in strength, the nonwetting character falls drastically, and is easily be penetrated by corrosive liquids. Oxidation of carbon is mainly due to gaseous oxygen (from air) but may also occur due to liquid phase, mainly slag. Oxidizing components present in slag react with carbon present in refractory and oxidize it. Slag oxidation occurs mainly due to the presence of FeO. An increasing amount of iron in slag drastically degrades the quality of the refractory and causes high wear. The presence of manganese oxide also increases the wear rate.

Oxidation of MgO–C refractory by gaseous components is mainly due to oxygen and carbon dioxide. The gaseous oxidation is strong when the hot refractories are exposed to open atmosphere (in between two heats). Gases directly react with the carbon present in the refractory and form its oxides. During the charging and discharging of molten metal and inspection or repair stage of refractories, refractories are highly susceptible to gaseous oxidation, as they are exposed to open air for prolonged time at high temperatures. The introduction of slag splashing technique after metal tapping (from converter) in steel-making process helps to prevent or reduce such oxidation. Slag splashing helps to coat the refractory by a thin layer of liquid slag and hinders the gaseous components to come in contact with the refractory. But when slag-coated refractory is cooled (mainly during repair) below the slag-freezing / solidifying temperature (~1000°C), the slag coating will solidify and shrink, causing crack and peeling off of the coat, and exposing the refractory to open atmosphere. This causes a sudden and huge oxidation. Figure 11.6 shows the beneficial effects of slag splashing and the effect of cooling of the MgO–C refractory on the oxidation behavior of the MgO–C refractory.

Carbon also gets oxidized by magnesia. Though MgO–C refractory is considered as a composite refractory, still there is some reaction between these components at high temperatures. Magnesia reacts with carbon and gets reduced to pure magnesium (in gaseous state at high temperature), and carbon converts to its oxides. This reaction may occur even within the interior of the refractory at high temperatures, and the product gases come out to the surface through the pores and the porous structure of the oxidized refractory. Magnesium vapor, at the surface of the refractory, reacts with the available oxygen from the atmosphere and converts to magnesia again.

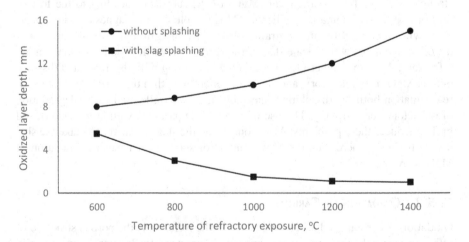

FIGURE 11.6 Effect of slag splashing on the oxidized layer depth of MgO–C refractory.

Thus, a newly formed magnesia layer covers the hot face of the refractory, reduces the surface porosity, and thus prevents the ingress of oxygen and further oxidation. This also helps in improving the corrosion resistance, as nascent magnesia layer on the refractory surface restricts the penetration of corroding slag.

$$MgO\ (s) + C\ (s) = Mg\ (g) + CO\ (g)$$

$$Mg\ (g) + \tfrac{1}{2}\ O_2 = MgO\ (s)$$

However, the interior structure of the refractory will become porous due to such reactions and will result in degraded properties. This reaction between MgO and C can be controlled by increasing the crystal size of the MgO grain, by using less-reactive fused MgO, and also by reducing the impurities of magnesia and increasing the lime–silica ratio.

11.5.3 ABRASION BY MOLTEN STEEL AND SLAG

Degradation of MgO–C refractory is also caused by the abrasion of the molten steel, especially in those areas where molten metal and slag has greater movement or impact. Physical wear due to mechanical action along with chemical dissolution due to the presence of moving fresh slag are the main reasons for this degradation. This can be controlled by dense and strong refractories, use of metallic additives, and lower carbon-containing refractories.

11.5.4 DEGRADATION DUE TO THERMAL AND MECHANICAL SPALLING

MgO–C refractories are also susceptible to thermal and mechanical wear due to the stringent application conditions. Repeated use of the refractory in steel converter causes thermal and mechanical spalling and fatigue effect, deteriorating the properties, Use of strong and dense refractories with proper carbon content, and proper design and arrangement of refractories are required to reduce this type of degradation.

11.6 MAIN APPLICATION AREAS

MgO–C refractories are only used in steel manufacturing process and to be specific mainly in the steel-making BOF (basic oxygen furnace) converters, steel ladles, and electric arc furnaces (EAF). Figure 11.7 shows the different areas of converter and ladle where MgO–C refractories are used. The major differences in quality of refractory used in the different application areas are their composition and associated properties.

For BOF working lining, mainly affected by slag and metal corrosion, MgO–C refractory with about 18% carbon is used. Higher carbon reduces the strength, but for such applications corrosion is the most important not the strength. But strength is important for the charge pad of BOF where the liquid metal and all other steel-making batch materials fall from a height. MgO–C refractory with about 9%–10% C is used for the charge pad and bottom applications. Once the charging is completed, these refractories will be covered under molten metal bath, and hardly there will be any

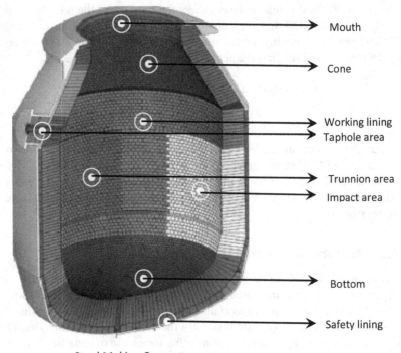

Mouth

Cone

Working lining
Taphole area

Trunnion area
Impact area

Bottom

Safety lining

Steel Making Converter

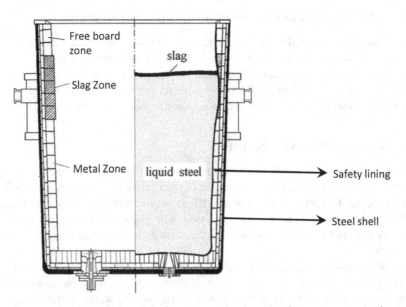

Free board zone

slag

Slag Zone

Metal Zone   liquid steel

Safety lining

Steel shell

Steel ladle (left side before use and right side after use, corrosion)

FIGURE 11.7   Details of steel converter and steel ladle for application of MgO–C refractory.

slag corrosion. So to resist the impact of falling batch materials, low-carbon MgO–C refractories are used. The whole BOF is hanging through metallic supports, and effect of stresses due to the hanging condition is mostly in the trunnion area. Also, the stresses increase due movement of the BOF during steel-making process. So refractories in these areas require strength along with flexibility, and about 12%–15% carbon containing MgO–C compositions are used for such applications. Similarly for taphole area, the refractories are under the tremendous movement of molten metal during tapping, so both the corrosion and strength are important and about 12%–13% carbon-containing relatively strong refractories are used for these applications. The cone portion (top) of the BOF is not in direct and continuous contact slag, so low-carbon-containing (around 8 wt%) MgO–C refractories are used for cone area.

For EAF slag zone, hot spot, and tap hole areas MgO–C refractories with 10%–12% C are used. Also, about 8% carbon-containing relatively strong refractories are important for EAF and for secondary steel making like VOD and VAD processes (vacuum arc degassing).

For steel ladle slag zone areas, MgO–C refractories with about 12% carbon are used, whereas for the metal zone, as the corrosion effect is much reduced, about 6%–8% carbon-containing refractories are used. Again, in the ladle bottom, where high strength is required for the impact of metal fall, about 5% carbon-containing refractories are used. But for the top part of the ladle (free board zone, not in contact with molten metal or slag) where the temperature, strength, and corrosion are not so critical, low-carbon (~5%) and low-strength refractories are used.

SUMMARY OF THE CHAPTER

Magnesia–carbon refractories are considered as a composite refractory where the good properties of magnesia, like basic slag resistance, are clubbed with the benefits of carbon-like nonwetting character and thermal shock resistances.

Use of carbon in developing magnesia–carbon refractories has started as impregnation in sintered dolomite and magnesia refractory, and later use of carbon as a batch component in the form of graphite came into the practice.

High purity magnesia aggregates (fused and sintered), graphite as carbon source, metal powders as additive (anti-oxidant) and resin as organic binder are used in making MgO-C refractories. These refractories are not fired, but tempered/cured, so pressing and curing are critical steps for proper property development in the refractory.

Increase in carbon content not only enhances the nonwetting character and thermal shock properties, but also increases the chances of oxidation, resulting in a porous and weak structure, easily corroded and penetrated by slag. A maximum of 20 wt% carbon is optimum for the performance of the refractory,

Degradation of MgO-C refractory occurs due to different factors during use, like dissolution of magnesia in slag, oxidation of carbon, abrasion of molten metal and slag, mechanical and thermal spalling, etc.

These refractories are mainly used in steel-making converter (BOF), EAF, steel ladles, and secondary steel making. The complete BOF lining is done by these refractories with varying carbon content due to stringent application conditions.

QUESTIONS AND ASSIGNMENTS

1. Why is carbon used in combination with magnesia?
2. Describe the selection criteria for magnesia and carbon for manufacturing MgO–C refractories.
3. What is antioxidant? Describe its role in the performance of MgO–C refractory.
4. Discuss the advantages and disadvantages of different antioxidants used in MgO–C refractory.
5. Why is organic binder used in MgO–C refractory? Why is resin a preferred binder?
6. Briefly describe the manufacturing technique of MgO–C refractory.
7. Describe the effect of increasing carbon content in MgO–C refractory?
8. What is the maximum carbon amount used in MgO–C refractory and why?
9. Describe in detail the how MgO–C refractories degrade.
10. Describe the different ways of oxidation of carbon in MgO–C refractory.
11. Describe the applications of MgO–C refractories.
12. Describe in detail with the reasoning for the use of different MgO–C refractories in steel converters.
13. Describe how one can improve the oxidation behavior of MgO–C refractory.

BIBLIOGRAPHY

1. Refractories Handbook, The Technical Association of Refractories, Tokyo, 1998.
2. *Harbison Walker Handbook of Refractory Practice*, Hassell Street Press, New York, 2005.
3. C. G. Aneziris, J. Hubalkova and R Barabas, Microstructure evaluation of MgO–C refractories with TiO 2-and Al-additions, *Journal of the European Ceramic Society*, 27 [1] 73–78 (2007).
4. U. Klippel and C. G. Aneziris, Prospects of ceramic nanoparticles as additives for carbon-bonded MgO-C refractories, *Proceedings 49th International Colloquium on Refractories*, Aachen, Germany, pp. 6–9 (2006).
5. C. G. Aneziris and U. Klippel, Thermal shock behaviour of carbon bonded MgO-C refractories with inorganic micro-and nano-additions, *CFI. Ceramic Forum International*, 83 [10] E50–E52 (2006).
6. C. G. Aneziris, U. Klippel, W. Schärfl, V. Stein and Y. Li, Prospects of ceramic nanoparticles as additives for carbon-bonded MgO-C refractories, *International Journal of Applied Ceramic Technology* 4 [6] 481–489 (2007).
7. C. G. Aneziris, D. Borzov and J. Ulbricht, Magnesia-carbon bricks: a high-duty refractory material, *Interceram - Refractories Manual*, 52 22–27 (2003).
8. B. Brenzy, The microstructures and properties of magnesia-carbon refractories, *Key Engineering Materials*, 88 21–40 (1993).
9. M. Guo, Degradation mechanisms of magnesia-carbon refractories by high-alumina stainless steel slags under vacuum, *Ceramics International*, 33 [6] 1007–1018 (2007).
10. A.S. Gokce, C. Gurcan, S. Ozgen and S. Aydin, The effect of antioxidants on the oxidation behaviour of magnesia–carbon refractory bricks, *Ceramics International*, 34 [2] 323–330 (2008).
11. N.K. Ghosh, D.N. Ghosh and K.P. Jagannathan, Oxidation mechanism of MgO–C in the air at various temperatures, *British Ceramic Transactions*, 99 [3] 124–128 (2000).

12. S. Behera and R. Sarkar, Nano carbon containing low carbon magnesia carbon refractory: an overview, *Protection of Metals and Physical Chemistry of Surfaces*, 52 [3] 467–474 (2016).
13. S. Behera and R. Sarkar, Low carbon magnesia-carbon refractory: use of N220 nano carbon black, *International Journal of Applied Ceramic Technology*, 11 [6] 968–976 (2014).
14. S. K. Sadrnezhaad, S. Mahshid, B. Hashemi and Z. A. Nemati, Oxidation mechanism of C in MgO–C refractory bricks, *Journal of the American Ceramic Society*, 89 [4] 1308–1316 (2006).
15. N. K. Ghosh, K. P. Jagannathan and D. N Ghosh, Oxidation of magnesia-carbon refractories with the addition of aluminium and silicon in air, *Interceram*, 50 [3] 196–202 (2001).
16. M. Bag, S. Adak and R. Sarkar, Study on low carbon containing MgO-C refractory: use of nano carbon, *Ceramics International*, 38 (3) 2339–2346 (2012).
17. S. K. Sadrnezhaad, Z. A. Nemati, S. Mahshid, S. Hosseini and B. Hashemi, Effect of Al antioxidant on the rate of oxidation of carbon in MgO–C refractory, *Journal of the American Ceramic Society*, 90 [2] 509–515 (2007).
18. M. Bag, S. Adak and R. Sarkar, Nano carbon containing MgO-C refractory: effect of graphite content, *Ceramics International*, 38 (6) 4909–14 (2012).
19. S. Behera and R. Sarkar, Effect of different metal powder anti-oxidants on N220 nano carbon contains low carbon MgO-C refractory, *Ceramic International*, 42 [16] 18484–18494 (2016).
20. R. Kundu and R. Sarkar, MgO-C refractories: a detailed review of these irreplaceable refractories in steelmaking. *Interceram*, 70, 46–55 (2021).
21. http://www.magnesita.com.br/en/solucao-em-refratarios/siderurgia/convertedor-ld.

12 Special Refractories

There are many refractory items that are used industrially but with a much reduced volume. But they are also equally important, as they possess some special characters and are essentially important for certain specific applications and conditions. These materials can be termed, in general, as special refractories, and some of those special refractories are described here.

12.1 ZIRCON AND ZIRCONIA REFRACTORIES

Zircon ($ZrSiO_4$) is the silicate of zirconium, containing 67% of ZrO_2 and 33% SiO_2, used as refractory material directly and also used as a raw material for zirconia (ZrO_2). The main reason for these materials to become a refractory is the presence of zirconia in them, which is high melting, chemically inert, and imparts special mechanical properties in the refractories.

Due to its refractory character, zircon is being used and studied since long though its thermal stability, decomposition and melting character varies from sources of origin and scientific reports. Zircons always contain some amount of hafnium; the HfO_2–ZrO_2 ratio varies but is normally about 0.01–0.04. The presence of other impurities like iron and rare earth are also common, and also affect the decomposition and softening behavior of zircon.

Zircon has a wide range of applications as a ceramic and refractory material, like,

- as refractory, it is mainly used for the construction of glass tank furnaces and nozzles for iron and steel industries,
- as molds and cores in precision, investment casting,
- as protective coatings on steel-molding tools,
- as an opacifier in the glaze on ceramic and whiteware industries due to its high refractive index,
- as the principal precursor for the preparation of metallic zirconium and its compounds, like zirconia.

The wide range of applications of zircon is due to its excellent thermophysical properties, such as low thermal expansion, good thermal stability, and high corrosion resistance against glass melts, slag, and liquid metals and alloys. Zircon is one of the most chemically stable compounds, and mineral acids other than HF cannot attack zircon. Very aggressive reaction conditions are required to break down the strong bonding between the zirconium and silicon parts of the compound.

Zircon decomposes by a solid-state reaction, and chemically pure zircon decomposes in solid state at 1676°C±7°C to form a mixture of tetragonal zirconia and cristobalite. But the presence of impurities reduces the decomposition temperature, which may even start at 1285°C. Specifically, the more the impurities present in

DOI: 10.1201/9781003227854-12

zircon, the lower the onset temperature of dissociation. To reduce the degree of zircon degradation, impurities, like, iron, titanium, aluminum, and alkali must be minimum. Also, it is critical for prediction of the expected life time of zircon refractories in contact with silica-containing melts having different cations.

Zircon-based refractories are prepared by using beneficiated zircon aggregate or sand, fine-milled zircon, and a temporary binder; then pressing the shape to desired shape and size; and finally firing the shapes. The firing temperature must be lower than the decomposition temperature of the zircon. The shaping of critical shapes is done by ramming, vibrocasting, or a series of air hammering technique called impact pressing. The main feature of the zircon refractories is their acid slag resistance, excellent thermal shock resistance, and good mechanical strength with wear and erosion resistances.

Zircon refractories are used as ladle nozzles for steel pouring, tundish metering nozzles, in furnaces for melting aluminum, and as compounds and coatings. Zircon refractories with at least 63% zirconia and maximum 20% apparent porosity are used in glass tank furnace. It is also an ideal mold material and chills sand due to its low thermal expansion rate, high thermal conductivity, and nonwettability with molten metal. Zircon is also used in core and mold washes to improve surface finish.

Zirconia is mainly obtained by chemical processing of zircon or by carbothermal reduction (fusion) of zircon in an electric arc furnace. Also, zirconia has a natural source, called baddeleyite, commonly found in igneous rocks containing felspar, zircon, etc. Baddeleyite is chemically homogeneous, but it may contain impurities such as hafnium (Hf), titanium, (Ti), and iron (Fe). Zirconia is an important refractory material due to its excellent high-temperature properties and chemical inertness. But high cost and cracking tendency have restricted wide applications.

At ambient condition zirconia has a monoclinic structure that instantaneously changes to tetragonal form during heating around 1180°C and from tetragonal to cubic form around 2370°C. On further heating, it melts around 2715°C. During cooling, also cubic form converts to tetragonal at 2370°C, but tetragonal changes to monoclinic at 980°C with a huge volume expansion during cooling. This spontaneous expansion (about 5% in volume) during cooling (generally cooling means contraction in dimension) causes cracking and failure of the material. Hence, for any high-temperature application zirconia is to be stabilized, where additions of CaO/MgO/ CeO$_2$/Y$_2$O$_3$, etc. are done to stabilize the tetragonal phase (called partially stabilized) or cubic phase (called fully stabilized) zirconia. Control on this polymorphic changes (and associated cracking) helps to incorporate microcracks within the zirconia-containing body and results in extremely high tough ceramics, which is highly beneficial for structural applications.

Zirconia refractories have refractoriness above 2000°C and also have excellent chemical resistance to the action of melts, alkalis, and most of the acids. But due to the high cost, they have only used for very specific critical applications. Main applicatin areas are the metering nozzle in tundish (for continuous casting of steel), inserts in the bore area in the slide gate plates (in steel ladles); in combination with carbon it is used as a band in subentry nozzle (SEN) to resist the corrosive attack of mold powders during continuous casting, etc. Photographs of these applications are shown in Figure 12.1. Zirconia is mostly used as a component in a refractory composition mainly to improve mechanical, corrosion, and thermal shock properties. It is also

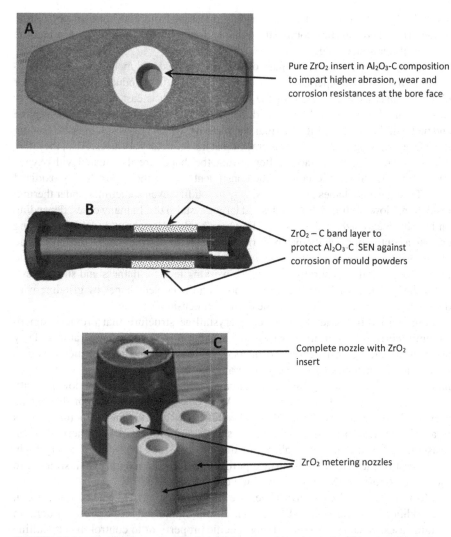

FIGURE 12.1 Applications of zirconia in refractory: (a) as insert in slide gate plate refractory, (b) as ZrO_2–C mix in slag band of subentry nozzles, and (c) as nozzle insert in tundish nozzle.

used as crucibles for melting platinum, palladium, and other metals and quartz glass, also in the construction of nuclear reactors. Lightweight zirconia products, fibers, and granular powders are suitable for high-temperature thermal insulation. Zirconia is also used as heating elements at temperatures up to 2200°C in furnaces with resistive and induction heating.

12.2 FUSED-CAST REFRACTORIES

Fused-cast refractories are a different class of refractories produced by melting the raw material of the desired composition, and then casting the molten mass in molds and solidified by cooling. They are widely different from conventional crystalline refractories of the similar composition due to superior corrosion resistance and the process of manufacturing.

The raw materials, generally pure oxides, are charged and melted, usually in electric arc furnaces, at temperatures above 2000°C. Sometimes induction, oxygen gas or plasma furnaces are also used for the melting. The melt is cast in sand, graphite, or cast-iron molds of desired shapes and dimensions. The main disadvantages of graphite and metal molds are their high thermal conductivity, which may result in faster cooling at surfaces, causing stress and stress cracking at the refractory surfaces. Also, they may discolor the cast shapes surfaces. After casting, the shapes are also treated with oxygen while in the fused state to convert the constituent ions to their most highly oxidized state. Then the cast shapes are cooled slowly to solidify, even sometimes under thermal insulation. Slow cooling results in desired large crystal sizes. In many cases, depending on the shape and size of the cast, cooling may be done for about 3–4 weeks. During cooling a contraction, cavity may be formed beneath the casting scar on the top surface of the shape and sawed to remove from the shape after cooling. The surface opposite to the casting scar is generally used as the working face for flatness and smoothness. A finished work of drilling or cutting is done on the cooled shapes by grinding with diamond tools to accurately match the desired dimensions.

Fused-cast refractories have a dense crystalline structure that provides superb strength, corrosion, and erosion resistances even at very high temperatures. They are quite stable against aggressive melts, such as glass and molten metal oxides. Fused-cast refractories have a very little amount of porosity, mostly closed pore in nature. The ingress of aggressive media cannot occur, because of the low porosity and the corrosion-resistant compositions. As a consequence, adhesion of slag can be prevented. Again, relatively high thermal conductivity (due to dense structure) results in a uniform heat distribution within the shapes and provides better thermal shock resistances. But the thermal stability of these refractories is usually not very high. Fused-cast refractories have a surface porosity of 1%– 3%, compression strength in the range of 400–700 MPa, and have high creep resistance.

Most of the fused-cast refractories are based on the Al_2O_3–SiO_2–ZrO_2 system. The addition of other oxides like calcia, chromia, and magnesia are also done in certain cases, especially to impart any specific property or to control the crystalline structure. Based on the mineralogical composition, fused-cast refractories are classified as below.

12.2.1 Fused-Cast Alumina Refractories

Fuse cast alumina refractories contain more than 90% Al_2O_3 and have excellent resistance to acid slag and chemical corrosion. According to the crystallographic analysis, fused-cast alumina refractories are commonly classified as alpha fused-cast alumina, beta fused-cast alumina, and alpha-beta fused-cast alumina. As per ASTM C 1547 – 02 (2013), these refractories are classified as per their soda (Na_2O) content, as obtained from chemical analysis and the beta (β) alumina ($NaAl_{11}O_{17}$) or beta" (β") alumina ($NaMg_2Al_{15}O_{25}$) phases content, as obtained from quantitative phase analysis (by x-ray diffraction) or image analysis study. Table 12.1 shows the

TABLE 12.1

Classification of Fused-Cast Alumina Refractories

Types	Soda Content (wt%)	Beta Alumina Content (vol%)
Fused α-Alumina	<1.1	<10
Fused α-β Alumina	3.5–4.7	50–65
Fused β-Alumina	5.2–7.7	>95
Fused β''-Alumina	4.5 (~8 MgO)	>95 as β'' ($NaMg_2Al_{15}O_{25}$)

TABLE 12.2

Classification of Fused-Cast Alumina–Zirconia–Silica (AZS) and Fused-Cast Zirconia Refractories

Types	Zirconia Content (wt%)	Zirconia Content (vol%)
AZS 21	19–23	18.5–22.5
AZS 33	31–34	30.5–33.5
AZS 36	34.5–37.5	34–37
AZS 40	38–41	37.5–41
Zirconia	>90	>90

details of the classification. Neutral character of fused-cast alumina against alkali vapors makes it the best choice for the downstream part of superstructures in glass tank furnaces.

12.2.2 Fused-Cast Alumina–Zirconia–Silica (AZS) Refractories

Fused-cast Al_2O_3–ZrO_2–SiO_2 refractories, abbreviated as AZS, are important for their corrosion resistance and wear properties mainly due to its compositions and considerable ZrO_2 content in them. They are the most widely used materials both in glass contact areas and superstructure of glass melting furnaces due to their very high resistances against glass and alkali vapor.

These refractories are further divided as per ZrO_2 content. According to ASTM C 1547 – 02 (2013), these refractories are classified by the monoclinic zirconia (ZrO_2) present, as determined by chemical analysis or quantitative image analysis. Table 12.2 shows the details of the classifications.

12.2.3 Fused-Cast Zirconia Refractories

The fused-cast zirconia refractories contain at least 80% of ZrO_2 and have excellent corrosion and wear resistance against molten glass. Therefore, it has been widely used in contact with molten glass in a glass melting furnaces.

TABLE 12.3
Classification of Fused-Cast Alumina–Silica Refractories

Types	Alumina Content (wt%)	Alumina:Silica Ratio
Mullite–Corundum	>67	>3.6
Mullite–Corundum–ZrO_2	>67 (>3 Zirconia)	>3.6

12.2.4 FUSED-CAST ALUMINA–ZIRCONIA REFRACTORIES

Fused-cast alumina–zirconia refractories are important for their high strength, excellent resistances against wear, and slag corrosion as well as long service life. They are mainly used in areas that require high abrasion and temperature resistance, such as gliding rail bricks in steel pusher metallurgical furnaces, the tapping platform style walking beam furnaces, etc.

12.2.5 FUSED-CAST ALUMINA–SILICA REFRACTORIES

Fused-cast alumina–silica refractories are characterized by excellent resistance to high temperatures, acid corrosion, erosion, and thermal shocks. They show high performance on reheating furnace beds, skid rails, and for the linings of industrial waste incinerators. It is widely used in the metallurgical and glass industries, ceramic kilns, cement industry, etc. ASTM C 1547 – 02 (2013) classifies (shown in Table 12.3) these refractories as per their alumina content and alumina to silica (Al_2O_3: SiO_2) ratio, as determined by chemical analysis using quantitative phase analysis or image analysis study.

Fused-cast refractories are the preferred refractories in contact with glass melts, due to their chemical stability, impermeability, and resistance to corrosion and erosion at the working temperatures of the glass melting furnaces. These refractories are used in glass melting and heating furnaces, petrochemical industries, and the most vulnerable sections of metallurgical industries, like oxygen-Bessemer converter linings. Powdered fused refractories are used to make critical parts and as fillers in linings of many furnaces, including induction furnaces.

12.3 INSULATING REFRACTORIES

Insulating refractories are thermal barriers for any high-temperature processing that do not allow the heat to escape from the desired processing and save energy. Any high-temperature processing demands for the maximum heat conservation so as to minimize heat losses for the maximum utilization of heat charged. This will result in minimum fuel consumption as well as high production as a result of for maintaining high working temperatures. As the cost of energy has increased, the role of insulating refractories has become more important.

The function of insulating refractory is to restrict or at least reduce the rate of heat flow out from the high-temperature process to open environment. Although it is not possible to completely prevent the loss of heat when a temperature gradient exists between two surfaces, but the use of insulating refractory can greatly reduce the loss and make the high temperature manufacturing process energy efficient.

Heat can be transferred from one place to another by three different mechanisms, namely, conduction, convection, and radiation. For solid materials, conductivity is the main mode of heat transfer. In conduction mode, heat is transferred by the transfer of energy from one atom to another (or molecule to molecule) in a material where atoms at higher temperature vibrate faster due to higher energy level and pass on the energy to the adjacent atoms present in a lower energy state (lower temperature).

So it is very common to design for an insulating refractory using a material that has a lower thermal conductivity. But in actual case the common refractories have thermal conductivity values in a very narrow range; it varies between 1 and 3 W/m.k. So only by changing the refractory material, effect on insulation will be marginal. To improve the thermal insulation character of refractories, air is incorporated as pores within the refractory body, as air has a very low thermal conductivity (~ 0.02 W/m.K). So a refractory having high porosity will have significantly low conductivity values due to the presence of air within it. Generally, the dense commercial refractories have surface porosity ~ 18%–24% and total porosity in the range of 30%–35%. Incorporation of air by creating porosity in insulating refractory increases this total porosity level to the tune of ~ 80%. But this increased insulation property is at the cost of lowering of strength, resistances against corrosion, abrasion, etc., properties. As the insulating refractories are having deteriorated mechanical, thermomechanical, corrosion resistance, etc., properties, they are not used at the hot face or in contact with any liquid at the high-temperature processing environment. Rather, they are used as a backup lining, behind the strong and dense refractory, only to retain the heat within the system and reduce the heat loss. So as the highest temperature that an insulating refractory face is the cooler side (lowest) temperature of the hot-face refractory, the temperature is much lower than the actual high-temperature processing. So the refractoriness and other thermomechanical properties required for insulating refractory are not that stringent as that important for hot-face refractories. Hence, generally cheap and commonly available fire clay-based insulating refractories are mostly used unless otherwise specified or required for any specific property requirement.

The structure of insulating refractory has minute pores filled up with air. The air inside the brick prevents the heat from being conducted, but the solid particles conduct the heat. So, to have the required insulation property in a brick along with other structural properties, a balance has to be made between the proportion of its solid particles and air spaces. Again higher the porosity, greater will be the insulating effect, but the size of the pore must be fine, and they must be uniformly distributed. As a bigger sized pore will have a greater volume of air, which again transfers the heat by convection mechanism within itself from the hotter side to the cooler side, and heat transfer will occur. Again, nonuniform distribution of porosity may result in the conduction of heat through the solid particles where porosity is less. Hence, evenly distributed, small-sized pores are desirable for the best insulating character.

Generally, the porosity within the refractory is intentionally created by the addition of fine organic materials to the mix, such as sawdust, straw, rice husk, etc., during mixing process of manufacturing. Then the mix is pressed, dried carefully, and fired. Organic materials are combustibles in nature and burn out during firing of the refractories creating internal pores. There are some special ways also to create porosity within the refractory, generally used to get a high level of porosity for special purposes only. These are:

1. use of special materials that expand and open up on firing;
2. use of volatile organic materials, like naphthalene, polystyrene, starch, etc.;
3. use of chemical bloating technique (like the combination of aluminum powder with NaOH solution);
4. use of porous-/open-textured materials like vermiculite, exfoliated mica, diatomite, insulating grog, etc.;
5. use of foaming agents, etc.

Insulating refractories can be classified by many aspects, like raw materials used, major oxide (constituent) present, and its content, porosity and strength level, etc. But the common and widely acceptable method of classification is based on their heat-withstanding capacity, as described below.

1. Insulating refractories that can withstand temperature up to 1000°C. In this category calcium silicate, siliceous earth materials, perlite, vermiculite, etc. are important.
2. Insulating refractories up to an application temperature of 1400°C. The common examples are lightweight fire clays, alumino silicates, lightweight castables, ceramic fibers, etc.
3. High-temperature insulating refractory that can withstand up to 1600°C. The examples are lightweight mullite and alumina, hollow sphere corundum (bubble alumina), alumina fibers, etc.
4. Ultra-high-temperature insulating refractory that can withstand up to 1800°C. Porous zirconia, nonoxide compounds, etc., are common examples.

The main advantages of insulating refractories are:

1. They reduce the heat losses through the furnace lining and make the high-temperature processes energy efficient.
2. They allow rapid heat-up and cooling of the furnace, and lowers heat capacity of the lining.
3. Thinner refractory lining can be designed with desirable thermal profile.
4. Reduction in furnace weight and thermal mass due to lower density.

But there are many disadvantages also, given as below:

1. Insulating refractories are poor in strength due to high porosity, causing a problem in structural design.
2. They are very weak in chemical resistance due to the porous structure, causing penetration of gasses, fumes, liquids such as slags, molten glass, etc., at high temperatures.

3. These refractories are weak against thermal spalling, as higher thermal gradient exists between the hot and cold surfaces (due to lower conductivity) and lower strength values.

The applications of insulating refractories range from laboratory furnaces to foundry furnaces and large tunnel kilns. Insulating refractories are very common as backup lining in most of the high-temperature operations where strength, abrasion, and wear by aggressive slag and molten metal are not at all a concern. These are widely used in the crowns of glass furnaces, tunnel kilns, etc., applications.

12.4   CERAMIC FIBERS

Ceramic fibers are also a class of insulating materials that can be spun and fabricated into textiles/cloths, blankets, felts, boards, blocks, or any other desired shapes. Ceramic fibers have low thermal conductivity and heat storage, light weight, relatively resistant to thermal shocks, and are chemically stable. Fluffy nature of the material prevents heat loss to a great extent and allows a very high rate of heating and cooling of the furnace/kiln with high-energy efficiency.

The main advantages of the ceramic fibers are as follows:

1. Better fuel economy due to low thermal mass reduced heat storage and better heat insulation.
2. Higher productivity for the user industry due to faster heating and cooling sequences.
3. Longer lining life with reduced maintenance costs, and
4. Ease of installation.

The main disadvantages of these materials are as follows:

1. Shrinkage at high temperatures and even on prolonging heating, causing open up (gap) of the refractory lining.
2. Poor mechanical strength (requires proper support for any design and structural use).
3. Sagging at high temperature due to softening tendency if proper support is not provided.
4. Cannot withstand any presence of liquids like slag, glass, and molten metals.
5. Expensive than conventional refractories.

The most common ceramic fiber systems are Al_2O_3, Al_2O_3-SiO_2, and ZrO_2. Ceramic fibers are made by blending the purer variety of raw materials followed by melting. Melting temperature may vary between 1800°C and 2400°C depending upon the composition. Molten stream is then broken by blowing compressed air or dropping the melt on the spinning disk to form loose or bulk ceramic fiber. The bulk fiber is then converted to various shapes like blanket, strips, veneering and anchored modules, paper, vacuum-formed boards and shapes, rope, wet felt, mastic cement, etc., for

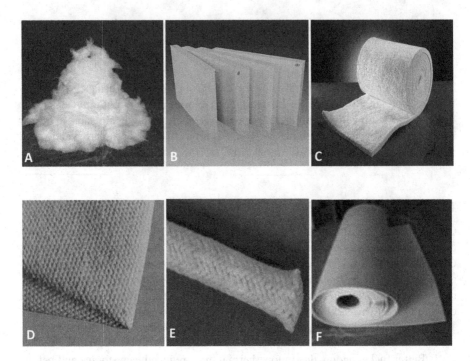

FIGURE 12.2 Different forms/shapes of ceramic fibers generally used in refractories: (a) bulk fiber, (b) board, (c) blanket, (d) cloth, (e) rope, and (f) paper.

various insulation applications. Different shapes made up of ceramic fibers, conventionally used for refractories, are shown in Figure 12.2.

Fibers with less than 60% Al_2O_3 are made from alumina, quartz, zircon, etc., and are directly extracted from the electrically fused melts. The melt stream is converted to the fiber by using oppositely moving quick rotating disks or by high-speed air jet blowing sideways (parallel) to the molten stream. These fibers have a glassy structure due to faster cooling. Again, fibers with more than 60% alumina are extracted from spinning solutions containing aluminum and silicon with some organic carrier (melting process requires much higher temperature and energy for processing). The spinner materials are then heat treated to remove the organic material first and then further heat treated to convert the fibers into solid, stable state.

Due to prolong service or excessive temperatures the structure of the fibers change and the glassy fibers convert to mullite and cristobalite. This is associated with changes in volume and strength. Originally the smooth surfaced fiber becomes rough; crystal appears, and the fibers become brittle. From the shrinkage parameter of the fibers, due to the structural change, the fibers are classified for its maximum permissible application temperature. The maximum temperature that a fiber can withstand without shrinking 4% linearly when heated from all the sides for 24 hours is called the classification temperature. Obviously, a higher alumina containing fiber is having a higher classification temperature and can be applied at higher temperatures.

Ceramic fibers are widely applicable for metal-treating furnaces, ceramic kilns, and many periodic heating operations. Fibers are common in expansion joints, door seals, preventing any heat leakage in any high-temperature applications. These materials can be directly used as hot-face refractory where liquid corrosion or penetration and strength requirements are not the major requirements. 52% Al_2O_3 and 48% SiO_2 containing fibers and shapes made from it are important as hot-face insulation up to 1400°C. Higher the alumina content, higher is the application temperature. Pure Al_2O_3 and ZrO_2 fibers, and their shapes are used even up to 1800°C.

12.5 CARBON REFRACTORIES

Carbon refractories behave differently than the typical oxide refractories, as carbon is a conductive material rather than insulating. They essentially consist of elemental carbon as the major constituent. The application of these refractories must be such that presence of oxygen or air will be minimum, otherwise the carbon will oxidize and burn out. Many a time's carbon and graphite are used to represent the same material, but precisely carbon is a general form that consists many of carbon-bearing sources in which graphite is also one natural source.

The main advantages of carbon as refractory are:

1. high refractory character with thermal stability in nonoxidizing conditions
2. poor wettability by polar liquids, in particular, by silicate-based slags
3. potentially high thermal conductivity
4. low thermal expansion properties
5. excellent thermal shock characteristics

In nature, carbon is available in several forms; the most common ones are diamond, graphite, and amorphous carbon. Diamond is a precious stone and insulating in character. Graphite is a crystalline, black, hexagonal layer network structured very soft material, conducts heat and electricity, and shows lubricity. Amorphous carbon has hexagonal layers as graphite but with disordered stacking, and results in amorphous character; but it is a hard material. Coke and charcoal do not have distinct crystalline character, and have properties in between of graphite and diamond. Varieties of properties coming from the same carbon material are due to the variation of carbon atom arrangements and its nature of bonding.

The raw materials used for making carbon refractories are petroleum coke, natural, and synthetic graphite (both the crystalline and amorphous forms), pitch coke, metallurgical coke, calcined anthracite, etc. These materials act as fillers for the carbon refractory and pitch, tar, resin, etc. are used as binders. Metal additives are also used as antioxidants, and also some other additives are added to enhance certain properties.

Among different sources, graphite is preferred by the refractory industries mainly due to its greater oxidation resistance, coming from its low porosity level. Oxidation can only occur through the actives sites in the granular planes. Also, flakiness (aspect ratio) is high for graphite, and so flaky graphite is preferred for its high thermal conductivity. This helps in further higher thermal shock resistances. But higher flakiness enhances the chances of lamination and anisotropic character. There is also an amorphous variety of graphite, which is found in nature in massive form. But natural graphite contains ash (about 8%–15%), which drastically degrades the properties. The ash contains about 45%–50% silica, 15%–20% alumina, 15%–25% iron oxides, little alkalis, and alkaline earth oxides. Natural graphite is required to be cleaned, washed, and purified (using flotation process) to reduce the impurities' content before use. Synthetic graphite is also used for making carbon refractories, generally produced from the used graphite electrodes that are crushed to get the desired fractions. Synthetic graphite is purer and has higher thermal conductivity, but dissolves easily in molten iron.

Petroleum coke is the calcined form of the green coke obtained from the petroleum industries. Crude petroleum is decomposed and distilled to separate out the precious gasses, oils, and other petroleum products, and the residue remained is the green coke. This green coke is calcined at about 1200°C to produce the petroleum coke. Also, the pitch coke is obtained in the same manner. The main advantage of these cokes is that they have very low ash content. Metallurgical coke of foundry variety is preferred for refractories due to its higher mechanical properties and abrasion resistances.

Anthracite is also used as a major ingredient for carbon refractories after calcination. Natural anthracite contains a good amount of volatile matters and ash content. Calcination can be done by gas firing or electrical firing. Gas-fired calcination process is done around 1200°C, and the calcined anthracite has better mechanical properties but poor thermal and shrinkage characteristics. Electrical calcination is done in a wide range of temperatures, varying from 1200°C to 2500°C. The calcination temperature affects the properties of the product, likes volume stability, electrical and thermal conductivity, strength, abrasion and alkali resistances, etc.

Raw materials as per desired size fractions are mixed with hot tar or pitch, and the mixing temperature is maintained about 50°C higher than the softening temperature of the binder, for better flowability and mixing. Any lump formation is avoided for a uniform texture of the mix. Mixed paste is then shaped by extrusion or by pouring into molds. Extrusion is done around 100°C–120°C and used mainly for the bulk production of dense shapes (low porosity) even at a lower binder content. There are chances of aggregate alignment in the extrusion direction and anisotropic character in the shaped products. Molding is done around 70°C–100°C and gives a relatively isotropic character in the shapes. Shapes can be rammed for better compaction in the hot condition and finally allowed to cool. Green shapes are then fired (baked) in the absence of air. They are first packed in coke to avoid oxidation and deformation due to softening on heating. This baking is done around 1300°C with a slow heating and cooling schedule, heating for 2–3 weeks and cooling for about a week. For graphite-containing compositions, the temperature used is much higher than the simple baking temperature. For compositions containing synthetic graphite, the firing temperature is further higher, and a graphitization process (at ~2500°C) is followed by baking to increase the strength and thermal conductivity of the products. Baked or graphitized refractories are finally cut and machined to match the required dimensions.

The properties of carbon refractories vary widely, depending upon the raw materials used for making them. Bulk density varies between 1.5 and 1.8 g/cc, and apparent porosity in the range of 12%–25%. They all have poor oxidation resistance but very high thermal stability, if oxidation is controlled. The cold strength varies between 40 and 100 MPa, and linear reheat change is negative within 1% at 1500°C (higher values for metallurgical coke). RUL values under 0.2 MPa load is above 1750°C, if oxidation can be restricted. Abrasion resistance is again dependent on the raw materials used. Graphite-containing compositions have higher thermal conductivity, but are softer and results in high abrasion loss compared to coke-based products. Carbon refractory based on electrographite are having about 25 times higher thermal conductivity than metallurgical coke-based ones at 200°C, and about 10 times higher at 800°C. Bonding characteristics and flakiness of graphite have resulted in such a high thermal conductivity values. Carbon refractories are weak against oxidation, which is caused by oxygen and carbon dioxide present in the environment, and also by the moisture present (leaking coolers). Carbon refractories are hardly attacked by molten slag and metal, as they have a nonwetting character due to strong covalent bonding among the atoms. The presence of alkalis is detrimental to carbon, as they may react and form intercalation compounds (say C16K, C24K, etc.). These compounds formations are associated with growth and expansion, and result in disintegration of the refractory.

The blast furnace is the largest user of carbon refractories followed by aluminum industries. Anthracite- and graphite-based carbon refractories are mainly useful for the hearth and bosh portion of the blast furnace. Carbon refractories are also important for the lower stack—runners of a blast furnace. For blast furnace bosh area, high thermal conductivity with abrasion resistance are required, Metallurgical coke-based products show good abrasion resistance, but they have low thermal conductivity. Whereas, electrographite-based refractories have very high thermal conductivity but poor abrasion resistance. For blast furnace applications, high resistance against abrasion is essential, so softer refractories are not preferred; however, formation of slag coating on the refractory surface reduces the abrasion effect on the refractory, and softer ones also perform well. This slag coating forms easily for refractories with high thermal conductivity. In aluminum industries, petroleum coke and amorphous carbon-based refractories are useful as anodes, cathodes, sidewalls, and hearth of the smelters. Graphite refractories are also important for various ferroalloys industries like Fe–Si, Fe–Cr, Fe–Mo, etc. Graphite–alumina and graphite–aluminosilicate refractories are important for various areas of steel and nonferrous metal industries.

12.6 SILICON CARBIDE REFRACTORIES

Silicon carbide (SiC) refractories are important, as they exhibit a unique combination of properties, like high strength both at ambient and elevated temperatures, high thermal conductivity, and high resistances against thermal shock, abrasion, erosion and corrosion against oxidizing slags and metals. SiC is prepared by the well-known Acheson process (about 125 years old) where quartz and coke are intimately mixed and heated for days in a special electric resistance furnace. Carbothermal reduction of silica ($SiO_2 + 3C = SiC + 2CO$) occurs at about 1600°C–2300°C. The process of making SiC and the furnace design have remained nearly unchanged since its inception in 1893. In a typical commercial process about 150 ton of batch material is used (mix of quartz and coke), and about 150 MWh of electrical energy is required to produce about 15 ton of SiC, and the remaining batch materials were are again reused in subsequent processes. SiC chunks obtained from the process are crushed, ground and graded as per the desired fractions for different uses.

Silicon carbide is a light weight, a very hard abrasion-resistant material with thermal conductivity ~21 W/mK at 1000°C. It is resistant to dilute acids and alkalis, but decomposes in fused alkali carbonates, borax, cryolite, etc. Also, SiC dissociates to Si and C in molten iron and steel, and oxidizes in the air above 800°C but is stable up to 2200°C in reducing atmosphere. Use of SiC has started as an abrasive material due to its excellent hardness, and, as on today, it is the second-largest used abrasive material. Grains and Powders are important as polishing, lapping, and shot-blasting material; it is also used for grinding wheels, finishing stones, coated abrasives like fiber disk, etc. They are also used as metallurgical additives as inoculants for cast iron, deoxidizer for steel. Sintered shapes are also used as electrical heating elements, mechanical seal, lightning arrestor, ignitor, etc.

Use of silicon carbide in refractories has started about the 1970s, since its use in a blast furnace. SiC is a strong, covalently bonded solid, and the main advantages of SiC as a refractory are:

- Light weight (bulk density <3 g/cc)
- High strength at high temperatures (up to 1600°C)
- Excellent thermal stability and thermal shock resistance
- High thermal conductivity and low electrical conductivity
- Resistance to oxidation, corrosion, abrasion, and erosion
- Resistance to molten nonferrous metals and slags

As SiC is a very strongly bonded material (about 89% covalent bonding), it has weak diffusivity and is difficult to get sintered. Hence, for masking any shaped product of SiC, very high temperature and pressure are required or a secondary material is to be added as a sintering aid. Again at high temperatures, the surface of SiC is covered with an incipient layer of silica (due to oxidation), different bond formations can be designed, for getting better sintered products, by the addition of proper additive utilizing this silica layer. But the final properties of the sintered SiC products are dependent on the bond types and are detailed in Table 12.4. Different types of bond used in SiC refractories are as follows:

- Oxide Bonded [OB-SiC]
- Nitride Bonded [NB-SiC]
- Reaction {Self-Bonded} [RB-SiC]
- Direct Sintered [DS–SiC]

Oxide bonded (OB) SiC are based on silica or silicate phases formed during firing due to oxidative/glass forming reactions. Addition of clay/silica/mullite is done with SiC batch composition, and the mixture is shaped by pressing/vibro casting/ pneumatic ramming and then fired in an oxidizing atmosphere. Oxide ingredients react with the incipient silica layer formed on SiC surface and develops the bond.

TABLE 12.4

Variations in Properties of SiC Refractories with Different Bonding Systems

Sl.	Property	OB-SiC	NB-SiC	RB-SiC	DS-SiC
1	Bulk density, g/cc	2.3 – 2.5	2.5 – 2.7	2.7 – 3.1	3.0–3.1
2	Approximate porosity, %	15–25	15–18	<1	1–3
3	Hot MoR @ 1250°C, MPa	18–25	45–80	300–450	550–650
4	Thermal expansion coefficient, $\times 10^{-6}\,°C^{-1}$	5.3–5.9	4.0–5.0	4.3–4.6	4.0–4.2
5	Thermal conductivity, W/mK	8–12	15–30	90–115	90–100
6	Thermal shock resistance	Good	Very Good	Excellent	Excellent
7	Oxidation resistance	Fair	Very Good	Very Good	Excellent
8	Corrosion resistance	Fair	Very Good	Very Good	Excellent
9	Abrasion resistance	Good	Excellent	Very Good	Very Good

The amount and nature of bond, and the properties can be tailored to the amount and type of the additions done. Generally, the hot strength is strongly affected due to the presence of the liquid phase at high temperatures. Amount of SiC present in the final refractories is around 80%.

Nitride-bonded silicon carbide refractories is a general name that was given to nitride bond [NB-SiC], oxy-nitride bond [ONB-SiC] and SiAlON bond [SAB-SiC] containing products. As per the desired bond in the final ceramics silicon metal powder, fume silica and fine alumina powders are added to different fractions of SiC batch materials. Conventional shaping techniques are used, and firing is done in a nitrogen atmosphere at around 1550°C (close to melting point of metallic silicon). During firing, nitrogen gas reacts with silicon metal forming silicon nitride (Si_3N_4), and these nitrides get accommodated in the porous structure of the shape. Hence, for silicon metal-containing composition, the dimensional change is very marginal; however, there is a significant increase in the weight. This NB-SiC product has a fibrous matrix of silicon nitride (and small amounts of oxynitride) in which SiC grains are strongly held. In oxy-nitride-bonded material, formed silicon nitride further reacts with silica-forming SiON phase in the fired shapes. Whereas in the compositions containing both silicon powder and alumina, the formed Si_3N_4 reacts with alumina-forming SiAlON. In ONB and SAB-SiC-bonded products, a thin film of SiO_2 is always present on the surface of SiC grains. NB-SiC refractories have a minimum SiC content of 75%.

In reaction bonded (RB) SiC refractories, different fractions of SiC are mixed with fine graphite, pressed and fired in a vacuum capable or environment controlled furnace between 1500°C and 1650°C. Granular silicon metal is placed in contact with the shapes at the high temperatures that melts and wets the SiC shapes and wicks through the compacts. During this, an exothermic reaction occurs due to the reaction between silicon melt and graphite forming nascent secondary SiC phase that acts as a bond material in the shape. Firing conditions along with amount of free graphite and silicon metal melt to be provided are required to be precisely controlled to control the formation of reaction bonding within the shape. These refractories contain at least 90% SiC in its composition.

Direct sintered (DS) or self-bonded SiC refractories are prepared from a mixture of submicron-sized SiC powders with sintering aids like boron, aluminum. Organic binders and resin as a source of carbon are also used in the composition. The compositions are shaped by casting or pressing and fired in a controlled atmosphere vacuum furnace at ~2000°C–2200°C. These refractories have very high shrinkage (~20%) due to high sintering, good density, and excellent high strength properties and contain about 99% SiC in its composition.

Applications of SiC refractories are mainly in the blast furnace linings, especially in the tuyere band, bosh, belly, lower and middle stack, etc. Upper stack region is having a low temperature but is under high wear due to abrasion of the solid batch materials, and OB-SiC refractories are used. In lower stack and belly portion, along with abrasion and alkali vapor attack, higher temperatures are also important, and NB-SiC refractories are preferred. Further, increase in temperature in the lower stack and belly portion of blast furnace and also, increase in corrosion of refractories from metal and slag due to alkali and zinc attack, demand for better quality of refractory for these portions. Hence, improved quality SiAlON bonded SiC refractories are suitable for such critical applications..

NB-SiC refractories are also used for the lining of Hall cells in primary aluminum production through Hall–Heroult process where highly corrosive cryolite liquid is present. SiC refractories are also used in the secondary melting of aluminum, like walls of reverberatory re-melting furnaces, immersed radiant tubes, etc. SiC refractories (generally OB-SiC types) are also important as a backup lining for the hot-face carbon refractories in the tap hole region electric arc furnaces used in ferro alloy industries. They help to control the metal viscosity due to its high thermal conductivity and act as heat sink. For the higher abrasion and corrosion resistances, NB-SiC refractories are used as lining for the reverberatory holding furnaces for copper melting. Also, different shapes of OB-SiC refractories are used in a vertical fractional distillation column to purify zinc from impure natural sources.

SiC refractories are also used as kiln furniture in the insulator, sanitary ware, tiles, table ware, etc., industries in the shapes of batts, plates, tiles, pillars, saggers, etc. SiC refractories are important due to the high temperature stability, strength, creep resistances, high thermal conductivity, and thermal shock resistances. These refractories allow high ware (being fired) to kiln furniture weight ratio, improves the productivity and fuel efficiency, etc. Different bonding systems are used for these applications depending on the application demand and properties achieved. Figure 12.3 shows the applications of SiC-based shapes as kiln furniture.

12.7 OTHER NONOXIDES USED IN REFRACTORIES

Continuous development and improvement of metallurgical industries, especially the iron and steel-making technology, have demanded refractories to cope up with the advancements and provide the desired performances in further stringent environments, maintaining both the environmental and economic factors. Though refractory comes under the broad classification of traditional ceramics, these high-performance

FIGURE 12.3 Use of SiC as kiln furniture: (a) on the tunnel kiln car where products are being fired and (b) an assembly of kiln furniture to make rack/shelf using SiC support rods and plates/batts.

refractories need to have certain specific application oriented properties that cannot be obtained from the conventional compositions. Hence, the refractory researchers are always in search of newer materials that can provide very specific properties essential for a particular application. These new materials are also expected to enhance the performance and service life greatly. These special additives are mostly nonoxide in nature and are commercially used in a very small fraction (volume) to the total refractory required/used. Mostly, they have excellent structural or mechanical characteristics, and are mainly incorporated in refractories to improve the structural properties. Few of these nonoxides are discussed below.

12.7.1 Boron Carbide (B$_4$C)

Boron carbide is the hardest material after diamond, and so it shows excellent abrasion and wear resistance. It has a good corrosion resistance, especially against the acids. Generally, it is associated with some amount of free carbon, as secondary graphite phase, and has a purity in the range of 78%–85%. This free carbon affects the mechanical properties. Also, it is low in oxidation resistance compared to silicon carbide, and low fracture toughness limits its wide applications. In commercial production, boron carbide is manufactured reacting and fusing boric oxide with carbon in an electric arc furnace.

B$_4$C is a strong covalent-bonded material, so to get a high dense product hot pressing (or hot isostatic pressing) technique is used. But these special manufacturing techniques restrict the dimensions and complexity of shapes, and costly cutting and grinding (using diamond tools) are required. The interesting properties of B$_4$C are extreme hardness and low thermal conductivity resulting in high brittleness, poor thermal shock resistances; good thermal-neutron absorbance, high wear resistance, etc. Some of the properties are listed in Table 12.5.

It is used extensively as ballistic armor and blast nozzle, and is the primary choice for control rods and other nuclear applications, abrasive as lapping and ultrasonic cutting, armor materials, wire drawing dies, powder metal and ceramic-forming dies, thread guides, etc. As refractory B_4C is important as antioxidant in MgO-C refractory, as it oxidizes at much lower temperature compared to conventional metallic antioxidants and protects carbon. Also, it is important as a component in nozzles for flow control of hot aggressive liquids.

12.7.2 Tungsten Carbide (WC)

Tungsten carbide is a very hard, wear-resistance material mostly important for industrial machinery, cutting tools, abrasives, armors, etc. Generally, it contains high percentages of either cobalt or nickel as a second metallic phase. These ceramic materials behave like a metal and are also called "ceramic metals" or "cermets." Tungsten carbide is prepared by reaction of tungsten metal and carbon between 1400°C and 2000°C, or by using a fluidized bed process that reacts either tungsten metal or blue WO_3 with CO/CO_2 mixture and H_2 between 900°C and 1200°C. Shapes of pure tungsten carbide can be made as an advanced technical ceramic using a high-temperature hot, isostatic, pressing process. This material has very high hardness and wear resistance, and is mainly used as a cutting tool, grinding media, abrasive water jet nozzles; however, its weight limits its use in many applications. Some properties of sintered WC products are shown in Table 12.5.

Tungsten carbide compositions have outstanding physical and mechanical properties, good impact strength; outstanding dimensional stability; exceptional heat resistance and resistance to thermal shock; good cryogenic properties and oxidation resistances up to ~600°C; and even good tensile strength. In refractory, it is used to impart high wear and abrasion resistance for critical application areas.

12.7.3 Silicon Nitride (Si_3N_4)

Si_3N_4 has the strongest covalent bond properties next to silicon carbide. It is used as a high-temperature structural ceramic due to its superior heat resistance, strength, and hardness. It also offers excellent wear and corrosion resistance. Its high strength and toughness make it the material of choice for automotive and bearing applications.

TABLE 12.5

Properties of Some other Metal Carbides Used in Refractory Industries

Material	Density	Melting Point	Thermal Expansion Coefficient	Thermal Conductivity	Microhardness	Elastic Modulus
	g/cc	°C	10^{-6}°C^{-1}	W/mk	MPa	GPa
B_4C	2.52	2350	4.5	0.3	3340	1.5
WC	15.5	2720	3.8	0.07	1780	8.1
TiC	4.9	3100	7.7	25	3300	4.6
ZrC	6.9	3530	6.7	0.05	2930	3.5

Silicon nitride can be made by heating powdered silicon powder in a nitrogen atmosphere in the temperature range of 1300°C–1400°C.

$$3\ Si + 2\ N_2 \rightarrow Si_3N_4$$

Due to the attachment of nitrogen with silicon, the weight of the sample increases with time, and the formation completes with a soaking time of 7 hours. It is difficult to produce in bulk, and it dissociates into constituent ions on heating about 1850°C, much lower than its melting point. For shaped products of silicon nitride, bonding is done by adding sintering aids that promote liquid phase sintering and densify at lower temperatures but with a compromise of the high-temperature properties. Silicon nitride has long been used in high-temperature applications, and is capable of withstanding tremendous thermal shock and thermal gradients generated in hydrogen/oxygen rocket engines. Some of the properties of sintered silicon nitride are given in Table 12.6. Various types of silicon nitride-sintered shapes produced by different methods are useful for heat exchangers, rotors, nozzles, bearings, valves, chemical plant parts, engine components, and armor.

12.7.4 Aluminum Nitride (AlN)

Aluminum nitride is mainly important for the semiconductor industries and relatively newer material in the technical ceramics family. In its pure form, it is white and tan, or gray color indicates contaminations. It is a covalently bonded material with a hexagonal crystal structure and requires sintering aids even in hot pressing technique. Oxidation starts at the surface from about 700°C in the air, which forms alumina on the surface and protects the material up to about 1370°C, above which rapid oxidation occurs. But AlN is very stable in inert atmospheres. It is attacked slowly by mineral acids and strong alkalis, and also hydrolyzes slowly in water. The key features of aluminum nitride are high thermal conductivity, low thermal expansion, and also low electrical conductivity. Some of the properties of sintered AlN are given in Table 12.6. The main application areas are molten metal handling components, material-processing kiln furniture, heat sinks, substrates for electronic packages, etc.

TABLE 12.6

Properties of Some Metal Nitrides Used in Refractory Industries

Material	Density	Melting Point	Thermal Expansion Coefficient	Thermal Conductivity	Knoop Hardness	Elastic Modulus
	g/cc	°C	$10^{-6}°C^{-1}$	W/mk	Kg.mm^{-2}	GPa
Si$_3$N$_4$	3.18	1900	3.3	22	2200	300
AlN	3.25	2200	5	125	1170	308
BN	2.27	3000	0.7	35	150	8

12.7.5 BORON NITRIDE (BN)

Hexagonal boron nitride is a white chalky material and is often called "white graphite." It has poor mechanical properties. There are different varieties of boron nitrides available, and nearly all show outstanding high-temperature resistance (>2500°C) in inert atmospheres but oxidizes above 800°C in an air atmosphere. BN shapes are made by hot pressing technique and can be machined using standard carbide drills. Due to its crystal structure, BN is anisotropic electrically and mechanically. BN composites show excellent thermal shock resistance, corrosion resistance to molten metals and chemicals, strain and damage tolerance, wear resistance, and machinability. Some of the properties of sintered BN are shown in Table 12.6. BN is used as vacuum components, low-friction seals, various electronic parts, nuclear applications, and plasma arc insulators. It is also used as a high-temperature insulator and in combination with TiB_2 in much ferrous and aluminum metallurgical applications.

12.7.6 METAL BORIDES

Among different metal borides, titanium diboride (TiB_2) and zirconium diboride (ZrB_2) are important. Both the borides are prepared by the carbothermal reduction of the respective oxides in the presence of B_2O_3, which is an economic and traditional process. Above 1000°, B_2O_3 volatilizes and forms the borides as per the following equations:

$$ZrO_2 + B_2O_3 + C \rightarrow ZrB_2 + 5CO$$

$$TiO_2 + B_2O_3 + C \rightarrow TiB_2 + 5CO$$

Products of TiB_2 and ZrB_2 are commonly sintered by hot pressing technique under vacuum or inert atmosphere. But, at ambient pressure, it takes about 1900°C–2200°C in a controlled atmosphere in the presence of sintering additives like metals, carbon, and rare earth oxides. Both the borides are having high melting point, low electrical resistance, high hardness, high thermal conductivity, and excellent corrosion resistance up to 1100°C. They resistant to HCl and HF, but are weak and decompose in alkali metal hydrate. Some of the properties of these borides are shown in Table 12.7.

TABLE 12.7

Properties of Some Metal Borides Used in Refractory Industries

Material	Density	Melting Point	Thermal Expansion Coefficient	Thermal Conductivity	Knoop Hardness	Elastic Modulus
	g/cc	°C	$10^{-6}°C^{-1}$	W/m.k	Kg/mm^2	GPa
TiB_2	4.5	3225	8.1	60–120	2200	54
ZrB_2	6.08	3250	6.9	45–135	1170	35

They are widely used as hard tools and cutting tools, wire stretching mold, sand jetting nozzle, etc. Both of them are important as high-temperature refractories (2000°C–3000°C) when the atmosphere is not oxidizing in nature. Titanium diboride is used in metallurgical applications involving molten aluminum. It is also used for some limited wear applications, such as ballistic armor to stop large-diameter (>14.5 mm) projectiles. TiB_2 is also used as vaporization boat or vessel of vacuum deposited film (like Al, Cu, Cr, Ag, Au, Ge, TiN, etc.). It is also used as reinforcement particle for Al metal-based composites used in automobile industries, impact-resistance cutting tool, crucibles, armor, etc. ZrB_2 is used as the protection tube for continuous temperature measurement in metallurgical industries, the secondary heating electrode of the tundish, etc.

SUMMARY OF THE CHAPTER

Zircon and zirconia refractories are important for the presence of ZrO_2 in them, which imparts excellent resistances against corrosion, wear, and abrasion. Zircon is a natural mineral that decomposes at high temperatures. But, in the presence of certain impurities, it decomposes at much lower temperatures. Hence, the highest application temperature is limited. Zirconia is rarely available in pure form and has polymorphic transformations. It requires stabilizing additives to avoid sudden volume changes with temperature and stabilization. These refractories are mostly used in glass contact areas, and iron and steel industries.

Fused-cast refractories are manufactured in a different way compared to conventional sintering techniques. As they are shaped from a molten mass, the presence of porosity is very minimum and have large-sized grains. These impart very high corrosion resistance and are mainly important as liquid contact refractory, mostly in glass industries.

Retainment of heat in a process is mainly done by the insulating refractory. A dense refractory has a higher thermal conductivity, and intentionally small and uniformly distributed pores are created within the refractory mass to reduce the low thermal conductivity. These refractories are used nearly for all the high-temperature industries but mostly as backup of the lining. Ceramic fibers are also highly insulating in nature and may have different compositions according to the raw materials used. Mainly the composition of the fibers dictates the property developed, and the suitable application temperature and environment. Different shapes may be formed using the fibers as per requirement to prevent heat loss and sealing of the furnaces.

Carbon refractories are made from different natural and synthetic sources of carbon, and are mostly important for the high thermal conductivity and resistances against thermal shock and corrosion. Oxidation of any carbon-bearing material limits its applications.

Silicon carbide is a synthetically prepared material and important for its thermal conductivity, thermal shock, and corrosion resistances. It has greater oxidation resistance than carbon and is more widely used compared to carbon refractories.

There are many other nonoxides that are used as an additive to impart certain specific properties in refractory compositions. Among them boron carbide, aluminum nitride, boron nitride, metal borides are important. All of them are prepared synthetically and are very strongly bonded material. Mostly, they are used to impart wear and abrasion resistances.

QUESTIONS AND ASSIGNMENTS

1. Why zircon and zirconia refractories are important? Mention the application areas of these refractories with the reasoning for their applications.
2. Why zirconia requires stabilizing additives? Describe in details.
3. How fused-cast refractories are made?
4. What are the different types of fused-cast refractories used?
5. What are the different applications of fused-cast refractories?
6. Why insulation refractories are important?
7. Why small and uniform distribution of pores are important for the insulating properties?
8. Why the insulation refractories are not used as hot-face lining?
9. How ceramic fibers are made?
10. Describe different classification techniques for ceramic fibers.
11. Why is carbon important as refractory?
12. What are the different sources of carbon for making refractory? How is a carbon refractory made?
13. What are the application areas of carbon refractories?
14. What are the advantages of silicon carbide as refractory material?
15. Describe different bonding systems used for silicon carbide refractories.
16. Mention the applications of silicon carbide as a refractory material.
17. What are the different nitrides used in refractory? Describe any one of them.
18. What are the different borides may be used in refractory? Write details about the metals borides.

BIBLIOGRAPHY

1. *Refractories Handbook*, The Technical Association of Refractories, Tokyo, 1998.
2. *Harbison Walker Handbook of Refractory Practice*, Hassell Street Press, New York, 2005.
3. B. Brezny and R Engel, Evaluation of zircon brick for steel ladle slag lines, *American Ceramic Society Bulletin*, 63 [7] 880–883 (1984).
4. D. Urffer, The use of zircon in refractories for glass making, *Industrial Minerals*, 344 49–53 (1996).
5. G. F. Comstock, Some experiments with zircon and zirconia refractories, *Journal of the American Ceramic Society*, 16 [1] 12–35 (1933).
6. I. A. Lowe, J. Wosinski and G. Davis, Stabilizing distressed glass furnace melter crowns, *Ceramic Engineering and Science Proceedings*, 18 [1] 164–179 (1997).
7. R. Sarkar and A. Baskey, Decomposition and densification study of zircon with additives, *Interceram*, 60 [5] 308–311 (2011).
8. R. L. Bullard, P. C. Cheng and B. F. J. Schiefer, Long term casting with zirconia nozzles, *Electric Furnace Conference Proceedings*, Vol. 49, pp. 345–354, Toronto, Canada, 12–15 November (1991).
9. ASTM Standard C1547-02(2013), *Standard Classification for Fusion-Cast Refractory Blocks and Shapes*, American Society for Testing Materials, Pennsylvania, US (2013).

10. L. J. Manfredo and R. N. McNally, The corrosion resistance of high ZrO2 fusion-cast Al_2O_3-ZrO_2-SiO_2 glass refractories in soda lime glass, *Journal of Materials Science*, 19 [4] 1272–1276 (1984).

11. D. Au, S. Cockcroft and D. Maijer, Crack defect formation during manufacture of fused cast alumina refractories, *Metallurgical and Materials Transactions A*, 33 [7] 2053–2065 (2002).

12. E. Novak, *Refractory Engineering: Materials, Design and Construction*, 2nd Ed., Vulkan-Verlag, Essen, 2005.

13. A. Rand, A. S. Ahmed and V. P. S. Ramos, The role of carbon in refractories, *Tehran International Conference on Refractories*, Tehran, Iran, 4–6 May, 2004.

14. A. R. Chesti, *Refractories: Manufacture, Properties, and Applications*, Prentice-Hall of India, New Delhi, India, 1986.

15. E. Mohamed and E. Ewais, Carbon based refractories, *Journal of the Ceramic Society of Japan*, 112 [10] 517–532 (2004).

16. J. H. Chesters, *Refractories- Production and Properties*, Woodhead Publishing Ltd, Cambridge, 2006.

17. A. L. Shashi Mohan, Silicon carbide refractories: India 2020, *Proceedings of the 8th India International Refractories Congress (IREFCON 10)*, pp. 86–91, Kolkata, India (2010).

18. A. Lipp, K. A. Schwetz, and K. Hunold, Hexagonal boron nitride: fabrication, properties, and application, *Journal of the European Ceramic Society*, 5 [1] 3–9 (1989).

19. G. Zhang, M. Ando, T. Ohji and S. Kanzaki, High-performance boron nitride-containing composites by reaction synthesis for the applications in the steel industry, *International Journal of Applied Ceramic Technology*, 2 [2] 162–171 (2005).

20. H. Tanaka, Silicon carbide powder and sintered materials, *Journal of the Ceramic Society of Japan*, 119 [3] 218–233 (2011).

13 Unshaped (Monolithic) Refractories

13.1 INTRODUCTION AND ADVANTAGES OVER SHAPED REFRACTORIES

Conventionally, refractories are considered as industrial products primarily used for any high-temperature processing, especially for the core sector industries like iron and steel, glass, cement, nonferrous metallurgy, foundry, ceramic, petrochemicals, etc. Also, it was a traditional concept to consider the refractories as a product with definite shape and size, used to make furnace and kilns for firing. But this traditional concept has changed with time, as a good amount of refractories are unshaped in nature. Unshaped refractory is not a new material; probably, it was used along with the shaped refractory initially. Placing and fixing of any refractory shapes essentially require mortars that belongs to the unshaped refractory. Any repair work used for refractory lining is essentially an unshaped refractory or uses the concept of unshaped refractory. But in early days it was not considered and recorded as a separate class of refractory, as unshaped refractory, which is being considered today.

The first well-recorded information on unshaped refractory is available from literature that in 1856 H. S. C. Devile prepared a refractory crucible using a refractory concrete based on alumina aggregate and alumina cement. Plastic refractory was first commercialized by W. A. L. Schaeffer in 1914. Schaeffer founded a company, originally named Pliable Brick Company, later changed to Plibrico Company LLC, which started the business of unshaped refractory. Schaeffer recognized the fact that refractory lining for various industries will become complicated, as simple shapes would not be sufficient for lining them. If we look into the recorded history of unshaped refractory then plastic refractory was started in 1914; then comes the castable refractory. First information on castable is available as a patent filed in 1923, wherein alumina–silica–zirconia system was shaped by simple casting technique and then fused to densify. Next comes the gunning mass that was patented (filed) in 1951 for a silica-based patching mass using a spray gun. Then in the 1970s deflocculated castable system was developed that can be cast in the molds using vibration at a lower water requirements. Next comes the self-flowing castables in 1987, where low cement castable composition was reported to flow under its own weight in Japan. Then comes the shotcreting materials in 1989, which was used to coat a refractory surface in Japan.

Unshaped refractories (also called monolithic refractories) are unfired products, and do not have any definite shape and dimensions. They can have similar chemical properties and may show similar or better physical, mechanical, thermomechanical, and other refractory-related properties than the shaped ones. The term monolithic comes from two different Latin words, namely, "*mono*" means "single" and "*lithus*" means structure. So monolithics are those refractories that can make a refractory lining with a single structure, that is, without any joint.

A shaped refractory is a prefired one with a definite and specific shape and dimensions and having a homogeneous structure and properties. But, an unshaped refractory is a mixture of graded refractory aggregates, and fines homogeneously mixed with bond materials and property-enhancing additives in a predetermined fixed ratio. They are packed as loose materials and transferred to the user industries where mixing with fixed amount of liquid is done using intensive mixers and installed by special application techniques. As per international standard ISO 1927 (5) and European standard EN 1402-1, monolithics are defined as "mixtures which consist of an aggregate and a bond or bonds, prepared ready for use either directly in the condition in which they are supplied or after the addition of one or more suitable liquids, and which satisfy the requirements of refractories".

The main advantages of unshaped refractories are as follows,

1. It involves no shaping and no firing; hence, manufacturing of these materials are simpler, less process dependency, less polluting, and economic.
2. Any shape and any dimension of the refractory lining can be fabricated, which is not possible for shaped products. Any complicated shape with as many curvatures, slopes, contours, etc., and large dimensions can be made by unshaped refractories that are not possible by shaping processes, like pressing, extrusion, etc. due to technological and economic reasons.
3. The unshaped refractories have much greater corrosion/wear resistance compared to that of shaped ones, as the number of joints is less (or without joint), because the joints are a weak point for corrosion and wear.
4. Installation of shaped refractory lining is a skilled job and requires a prolonged time, whereas unshaped refractory placement is simple, less skilled, and quick. Thus, installation process is also easy and economic.
5. Unshaped refractory can bond with itself easily, even with a fired lining. Thus, when a lining is worn out due to use, the lost lining can be filled by the fresh material with similar composition and fired, the whole lining can act as afresh. But for brick lining, when it is worn sufficiently, the lining has to be dismantled, and rest of the brick has to be thrown away, and a completely new lining is required/done. Thus, nearly half of the brick remains unutilized every time, and complete relining is also a time consuming and costly affair.

Due to these various advantages, the unshaped refractory has replaced the shaped ones in a big way. Mostly in the advanced countries, the use of unshaped refractory is more than 50% of the total refractory. But for developing and under developed countries the usage is far below, mainly due to difficulties in certain process automation, control on installation and application techniques, and for control on the high-temperature processing at the user industry.

13.2 CLASSIFICATION

Unshaped refractories can be classified as per the main constituent present in them, like alumina based, magnesia based, carbon containing, etc. However, from the major constituent point of view, unshaped refractories are mostly based on alumina and alumino-silicate materials. Basic unshaped refractories are used but only for certain special applications and with specific application techniques. Silica-based unshaped refractories are also very limited. Carbon-containing unshaped refractories are also available but with limited use and very low carbon content. Carbon is hydrophobic in nature, and it does not disperse well in aqueous medium. So nonaqueous liquid is essential to disperse carbon, which are generally highly viscous in nature. Hence, homogeneous mixing with uniform properties is difficult to obtain for compositions with higher carbon content.

Again, as the unshaped refractories are not compacted (only the shape is given by different techniques), special binders are required for strength development, both at ambient and high temperatures, and also to retain the shape at low temperatures. There are different bonding materials, and classification is also done based on these bonding materials present in the unshaped refractory. For example, cement-bonded, gel-bonded materials; but they are not major classification of unshaped refractory and generally used as subclassifications.

Most common and widely used classification of unshaped refractories are based on the application techniques, and unshaped refractories are also called as per the application method used for installation. To cope up with a specific application technique, sometimes special additives are added, and also the properties may vary according to the installation methods. Table 13.1 provides the details about such classification. Among all these types of unshaped refractories, castable is the most studied, important, and commercially used in various industries. Though most of the unshaped refractories are important as repair mass or supporting the primary lining, castable is mainly used as primary refractory lining replacing the conventional shaped refractories. Advancement in castable technology has made it possible to replace the shaped refractories, even more than 70%, in many of the advanced countries.

13.3 SPECIAL RAW MATERIALS AND ADDITIVES

Among all the different types of unshaped refractories, the major constituent is mainly alumina and alumina-silicates, and in some limited cases, it is magnesia or silica. Detailed discussion on the commercially available and widely used raw materials for the above oxides are done in the earlier chapters for the individual shaped refractories. Mostly, these raw materials constitute the major portion of the different unshaped refractories. Other than these, matrix modifiers are added, which are not used in shaped refractories. Matrix modifiers are always in fine fractions and essentially required for the betterment of the properties. Other than the major constituent, there are different other essential components required to make unshaped refractory. These can be classified as binders and property-enhancing additives, which are detailed below.

TABLE 13.1

Application Techniques and the Names of the Unshaped Refractories

Application Technique	Name of Unshaped Refractory	Description
Casting/Pouring	Castable	A dry mix of different sized materials mixed with liquid (water) at the application site. Installation is done by simple pouring method in a gap space, made by using a former, where the casting/refractory lining is to be done, with or without vibration.
Ramming	Ramming mass	Granular material contains lesser fines than castable, placed by ramming methods using hand/pneumatic rammers and shovel tamping. It is mostly done for the bottom surface, and rammed material thickness is kept below 100–150 mm.
Gunning	Gunning mass	A dry (or semidry) mix of different fractions finer than castable. The dry mix needs to flow under pneumatic pressure through a hose pipe of gunning equipment from the stock hopper and generally mixed with liquid (water) at application nozzle.
Pasting	Plastic/pliable/patching mass	It is not a dry mass, rather a ready-to-use material, supplied as a premixed condition with liquid. Consistency and quality of the material vary with the specific application method. Conventionally used for small repairs (like cracks) but can also be applied for greater area repair by mallet hitting or pneumatic rammers.
Spaying/Shotcreting	Spray mass/shotcrete	These are sprayable refractory, similar to gunning in character but with finer sizes. Shotcrete is a further advanced form for better application and performance of the lining.
Vibration (dry)	Dry vibratable mass	Dry premixed mass generally contains thermo-setting binders are applied with vibration for uniform distribution and compaction. The materials sets and hardens on heating.

13.3.1 Major Constituent Fines

Unshaped refractories widely differ from shaped refractories, as they are not compacted during the shaping (shaping occurs during installation without load unlike shaped refractories), and so strength development during sintering is weak in these compositions. Also the concept of packing used for unshaped refractories is never to attain the most-dense structure, because they need to move, either flow or transported, during application/installation, and the highest packing (compaction) will restrict this behavior. Hence, the sintering in the composition during firing will be limited due to these less-packed conditions, which may result in poor densification and strength in the final product. These problems are somewhat taken care by addition of special fine fractions of the major constituents with special characteristics. These fine fractions need to help the flow properties, and also to improve the sintering and strength development. Thus, these fines modify the matrix part of the unshaped refractory and are called matrix modifiers. For alumina sources, some special types of property tailored alumina fines, which are generally not used in shaped refractories, are used in unshaped refractories to modify the matrix so that both the requirements can be met. These special aluminas are technical alumina fines and reactive alumina. Fineness and shape of these fine aluminas are important. Again, the particle size distribution is tailored for these special fine fractions, by using bimodal or multimodal distribution, so that they can improve the compaction of the whole composition. These alumina fines are essential for the matrix phase in technologically improved unshaped refractories. But for very conventional unshaped refractories, these matrix modifiers are not used mainly due to their high cost.

13.3.2   BINDER

Binders are essential for the development of strength at the ambient and elevated temperatures, and also to retain the shape at green conditions, as no compaction is involved in shaping process during installations of unshaped refractories. Also, binders help to sinter the compositions, as there is no compaction during shaping, mass transfer for sintering is poor for greater interparticle distance, resulting in poor densification and strength development. Initially, plastic clay was used as binder for unshaped refractory. Plastic nature helped to retain the shape, stickiness provided the strength at low temperatures, and fine size of clay particles has improved the densification behavior. But higher water requirement, shrinkage, and poor hot properties were the associated drawbacks. Then came the alumina-containing cement. During the initial days the alumina content of the refractory cement was less than 50%, and the developed properties were also poor. As the time progressed, the quality, properties, and performances of the alumina cements improved significantly with the increase in alumina content. But the presence of lime in the composition resulted in poor high-temperature properties, and refractory researchers, manufacturers, and users were looking for a binder without lime. Then came the cement-free bonding systems, like sol–gel, hydrated alumina, phosphates, etc. But, all the different bonding systems have their own limitations and are used in a limited manner. And, still today, high-alumina cements are used as the major bonding material for unshaped refractories.

13.3.2.1   High-Alumina Cement

At the very beginning era of unshaped refractories, people thought of using ordinary portland cement (OPC) as a binder for unshaped refractory due to the similarity of the material with concretes. But OPC has about 60%–65% of CaO, which is deadly for the refractory industries. Also due to dehydration of the bond, the structure will collapse above ~625°C and in the cyclic condition above even 400°C. So OPC is never used as a bond in unshaped refractory.

Alumina-based cement was first patented by Lafarge in 1908 in France, and the first commercial production was also started by Lafarge in 1913. But, originally alumina cement was developed and used for construction industries due to its better corrosion resistance than OPC. Use of alumina cement for high-temperature applications started in the mid-1920s, both in US and in France, when calcined clay, crushed fireclay bricks were mixed with bauxite-based alumina cement. But, these materials were of very low grade due to impurity, poor mixing, and application technique.

In the initial days, use of cement was high, ~ 12–18 wt%, and so the water requirement (for cement hydration and flow) was also very high, about 10–20 wt%. Hydration of cement results in significant strength development at low temperature. During firing, the hydraulic bond of cement particles breaks due to dehydration, removal of moisture occurs, resulting in a porous and very weak structures. So high density and strength were difficult to attain without liquid phase formation. Also, high amount of lime coming from cement easily reacts with the alumina, and aluminosilicates fines and forms the liquid phases at a much lower temperature and restricts high-temperature applications. Hence, these unshaped compositions were only applicable for low temperatures.

As the operating temperatures of the refractory user industries, especially the metallurgical sector, are increasing due to greater purity of their products, high-temperature withstanding refractories are in demand. To meet the requirement, unshaped refractories with reduced cement content were being tried since long. In 1969, the first patent was filed in France for the low-cement-containing composition, wherein the cement was reduced to 4–6 wt% (from 12 to 18 wt%) with near equal amount of water demand by using some flow-modifying additives, namely silica fume and dispersants. Fine silica fume particles improve the flow and packing of the unshaped refractory, as was done by cement and cement hydrates. This development has increased the application temperature from 1300°C–1350°C level to 1500°C–1550°C level. Further reduction in lime content by reducing the amount of alumina cement in the composition continued to reduce the liquid phase formation, thus improves the high-temperature properties. Reduction in cement content without compromising the properties has been made possible by controlling quality of aggregates and fines, their granulometry, quality of cements, use of additives, etc. Also, newer classes of cements containing lower lime (higher alumina) are developed and used in unshaped refractories, maintaining all other properties.

The alumina cement can be of different grades depending on the alumina content, and accordingly their raw materials, properties and application areas change. Low alumina-containing cements, initially developed and still in use, are prepared from lime stone and bauxite. Impurities of these raw materials remain within the cement composition and limits its application temperature to 1250°C–1300°C. Use of purer variety of natural raw materials marginally improves the properties and the application temperatures. Pure variety and high-alumina-containing cements are prepared mainly from hydrated lime (produced by hydration of calcined lime stone) and calcined alumina (from Bayer's process). These cements can have very high alumina content, even up to 80 wt% Al_2O_3, with very minimum amount of impurity. They are commonly referred to as high-alumina cement (HAC).

Calcium aluminate (high-alumina) cements are based on CaO–Al_2O_3 phase diagram, as shown in Figure 13.1. The two major and desirable phases present in purer variety of HAC are mono-calcium aluminate ($CaO. Al_2O_3$, CA)—the principal reactive

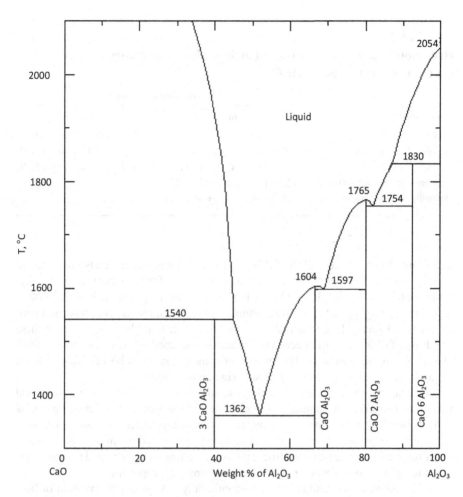

FIGURE 13.1   Phase diagram of CaO–Al$_2$O$_3$ system.

phase that hydrates and provides bond, and calcium di-aluminate (CaO.2Al$_2$O$_3$, CA$_2$), the secondary hydraulic phase. There may be some unwanted phases also like dodeca-calcium hepta-aluminate (12CaO.7Al$_2$O$_3$, C$_{12}$A$_7$), which is flash setting in character, and calcium hexa-aluminate (CaO.6Al$_2$O$_3$, CA$_6$), which is nonhydraulic in nature. Presence of 3CaO.Al$_2$O$_3$ (less likely to be present due to wide compositional differ-ence) and C$_{12}$A$_7$ make the cement very fast setting, or flash setting, due to very high amount of lime; whereas CA$_6$ makes it nonhydraulic due to minimal presence of lime. In impure cement system, there are some nonhydraulic impurity phases also, like, gehlenite (2CaO. Al$_2$O$_3$. SiO$_2$), calcium titanate (CaO.TiO$_2$), calcium ferrite (CaO. Fe$_2$O$_3$), and tetra calcium alumino ferrite (4CaO.Al$_2$O$_3$. Fe$_2$O$_3$), etc. Table 13.2 shows the hydration scheme of the major hydraulic phases present in calcium aluminate.

Metastable hydroxide CAH$_{10}$ and C$_2$AH$_8$ phases convert to stable C$_3$AH$_6$ phase during curing and drying of the unshaped refractory after shaping. During drying

TABLE 13.2

Hydration Reaction of Two Major Calcium Aluminate Phases (Where $C = CaO$, $A = Al_2O_3$ and $H = H_2O$)

Temperature	Hydration Reaction for	
	For CA	**For CA_2**
< 18°C	$CA + 10H = CAH_{10}$	$CA_2 + 13H = CAH_{10} + AH_3$
Between 18°C and 30°C	$2CA + 11H = C_2AH_8 + AH3$	$2CA_2 + 17H = C_2AH_8 + 3AH_3$
> 30°C	$3CA + 12H = C_3AH_6 + 2AH_3$	$3CA_2 + 21H = C_3AH_6 + 5AH_3$
Final changes (depends on time, humidity, and temperature)	$2CAH_{10} = C_2AH_8 + AH_3 + 9H$ $3C_2AH_8 = 2C_3AH_6 + AH_3 + 9H$	

and firing of the product, all the CAH_{10} and C_2AH_8 phases completely converted to C_3AH_6 phase below 200°C and, on further heating ~300°C, it converted to anhydrous metastable $C_{12}A_7$ phase. On further increase in temperature, between 900°C and 1000°C, $C_{12}A_7$ reacts with finer alumina particles in the matrix phase and converted to CA phase. Due to dehydration and hydraulic bond breaking activity there is a drastic fall in strength of cement-containing unshaped refractories above 300°C, and the fall is stronger when the amount of cement content is higher. This reduced strength remained till the starting of sintering above 1000°C.

The amount of water used in making the unshaped refractory plays an important role in the property development. Water is required for hydration of cement particles and also for flow consistency of the material. Any excess water will cause extra void space in the structure, and will result in reduced density and strength. Again, excess water helps to grow the cement hydrate phases, resulting in bigger sized hydrate crystals, bigger sized pores among them, and poor strength properties.

Again, when the cement contains predominantly CA_2 phase, the amount of lime is less, and the cement will have reduced hydraulic activity. But there are techniques, like increased fineness, increased reactivity of CA_2 phases by controlling grain growth, etc., by which the hydration tendency can be improved. The presence of higher alumina in CA_2 results in excess AH_3 formation, which fills the porous structure of the unshaped refractory in a better way, resulting in improved compaction and strength values. Also, the CA_2 phase has about 160°C higher decomposition temperature compared to that of CA phase (Figure 13.1), so CA_2 phase containing HAC are better in high-temperature properties also.

13.3.2.2 Colloidal Silica (Silica Sol)

Use of calcium aluminate cement results in relatively low high-temperature properties due to formation of low melting phases. For high cement-containing compositions, total CaO present is high enough that reacts with the impurity and constituent phases and forms low melting compounds in the $CaO–Al_2O_3–Fe_2O_3$, $CaO–Al_2O_3–SiO_2$, $CaO–Al_2O_3–SiO_2–TiO_2$, etc. systems. A reduction in cement content reduces the amount of CaO considerably in the system, reduced the amount of hydrated phases, results in reduced room temperature strength, fines content, and flow character. The addition of other fine materials, like silica fume, improves certain properties but may affect the high-temperature properties by forming liquid phases, mainly in the $CaO–Al_2O_3–SiO_2$ systems. Also, corrosion resistance of lime-containing system was not the best. Curing and dewatering steps also require special attention to avoid any explosion, spalling, and cracking. Hence, the refractory researchers were in search of CaO-free bond systems for use.

Among different lime-free bonds, colloidal silica or silica sol bonding has got the wide popularity and commercial success. Colloidal silica is a stable dispersion of silica particles in a liquid, usually water, where particles are small enough and suspended so that gravity doesn't settle them but large enough so that they do not pass through a membrane, and allow other molecules and ions to pass freely. Colloidal silica is denser than water due to presence of silica particles and electrostatically stabilized to form stable suspension. The basic units of colloidal silica are $[SiO_4]^{4-}$ tetrahedral, which are randomly distributed, resulting in an amorphous nature of the material.

Colloidal silica is prepared in a multistep process. An alkali-silicate (sodium silicate) solution is first neutralized to form silica nuclei, which are usually just a few nanometers in diameter. Polymerization of these nuclei starts without the alkali ions, when the stability of the sol is disturbed by change in pH, temperature, etc., and thus the sol particles grow in size. So to maintain the stable sol character controls on pH, temperature are essential, and, generally, the sols are stored between 5°C and 30°C. At very low temperatures the sol loses its stability and precipitates out as silica particle. Again, at elevated temperatures, the size of the particles increases and thus decreases the long-term stability of the silica sol. Other than the refractory applications, colloidal silica is also important for manufacturing of coatings, catalysts, paper industries, moisture absorbent, etc.

The use of silica sol replacing alumina cement is a great improvement in unshaped refractories and resulted in improved high-temperature properties. The principle behind this bonding is the formation of a "gel" structure from "sol," which surrounds and encapsulates the refractory aggregates through a three-dimensional skeleton network. During drying, the hydroxyl groups (Si–OH) of sol on the surface of the particles convert to siloxane bonds (Si–O–Si) due to removal of moisture and forms the rigid three-dimensional network. This gelation can also be induced by water removal, pH variations, using additives called gelling agents, etc. Figure 13.2 shows the schematic reaction of silica sol converting to siloxane bond (gel). This gelled

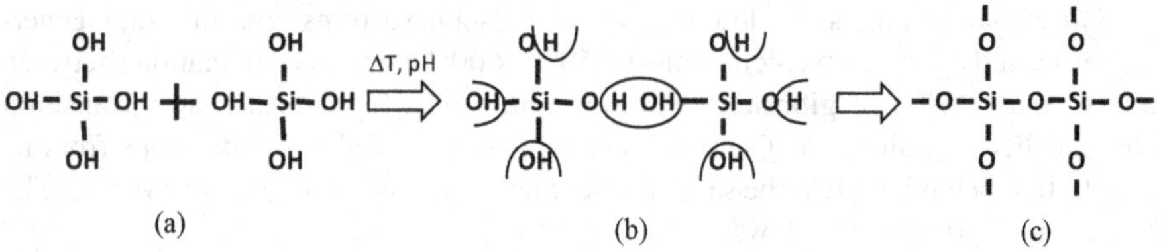

FIGURE 13.2 Schematic reaction of silica sol converting to siloxane bond (gel formation).

network around the refractory particles provides strength to the system after drying. This type of bonding is termed as coagulation bonding. After removal of moisture the structure is highly permeable and provides path for easy removal of any further moisture (chemically bonded water) from the unshaped refractory, reducing cracks and explosive spalling. On further heating the fine gel particles help in sintering by forming ceramic bonding at a lower temperature. Also, silica sol helps in ability to flow due to its fine size and spherical shape. This formation of mullite on firing in alumina-based compositions also helps in providing extra bond for strength development, improving the corrosion resistance and the hot-strength properties of the unshaped refractories.

The advantages of colloidal silica over high-alumina cement are listed below:

1. Less mixing time due to absence of additives
2. Higher viscosity (than water) improves separation between the refractory particles that improves flowability.
3. Reduced drying time and drying defects due to the absence of free water for mixing and permeable structure.
4. Inherent formation of mullite in alumina based compositions improves corrosion resistance and hot-strength properties.
5. Better high-temperature properties due to absence low melting compounds in $CaO–Al_2O_3–Fe_2O_3$ and $CaO–Al_2O_3–SiO_2$ systems.
6. Better high-temperature properties result in longer campaign life and reduced downtime of operation.
7. Longer shelf life due to the absence of any hygroscopic phases.
8. Fine silica particles from the sol coat the nonoxides (if present in composition) in a better way, resulting in improved oxidation resistance and better performance for nonoxide-containing compositions.

13.3.2.3 Hydratable Alumina

Hydratable alumina binders are increasingly used in no-cement unshaped refractories for its unique character. In this bonding, an inert matrix containing alumina formed from hydratable alumina binder can provide a nonreactive protective border surrounding alumina aggregates. Thus, a complete alumina system, having both the aggregate and matrix as alumina, without any secondary phase can be developed that will have improved slag resistance, fracture toughness, hot strength, and resistances against thermal shock and erosion. The absence of lime and other impurities nullifies any chance of low-melting phase formation, and high strength develops in the composition due to the ceramic bond formation at high temperatures.

Hydratable aluminas are low crystalline mesophase transition aluminas generally produced by vacuum calcinations (~550°C–600°C) or flash calcination (between 600°C and 900°C) of gibbsite, resulting mainly in a high-surface area transition phase, called rho-alumina. The bonding character of rho-alumina develops from its rehydration behavior when the same is in contact with moisture (or water vapor). The hydration reaction is as follows:

$$rho - Al_2O_3 + H_2O \rightarrow Al_2O_3.\ 3H_2O + Al_2O_3.\ (1-2)\ H_2O$$

A thick layer of gel is formed during hydration, which subsequently crystallizes partly to bayerite and boehmite phases, and the rest remaining as amorphous gel. Amount of the gel formed depends on the hydration temperature and pH. Interlocking bayerite crystals and the filling up of the pores and interfacial defects by the gel form a honeycomb type structure on the refractory aggregates that provides the green strength to the unshaped refractory. Such crystallization also favors the formation of crystals on the surface of the aggregates, connecting adjacent grains to the surrounding matrix. As the bond formation and strength development are based on the hydration of bond material, they form hydraulic bonds. During heating, the hydrated phases lose their chemical water and convert to stable α-Al_2O_3 fine particles, which help in sintering at higher temperatures. However, hydratable aluminas have few limitations, like they are highly susceptible to explosive spalling during drying due to less permeable structure, and also the development of strength at low temperatures are poor. Economically hydratable alumina is also not favorable.

13.3.2.4 Phosphates

Phosphate-bonded refractories are well known since 1950s, but wide commercial application has started late. This category comes under the broad classification of chemical-bonded refractories where phosphate acts as the chemical bond. Initially, the main chemical bond used for refractories is sodium silicate, but it is used in very limited applications due to its low melting point. Phosphate bonding is typically used for making "plastic" refractories, and the main features are that it has no lime, and it does not set easily at ambient conditions. Freedom from setting allows it to remain as moistened condition for nearly indefinite time periods and allows great working time for the lining under repair. But its application is also limited mainly due to its poor strength development. Poor strength, poor resistance against corrosion, and wear for phosphate-bonded materials are due to relatively weak and porous structure coming from higher liquid content for installation and application. In contrast to cement-bonded compositions, which perform better in neutral and basic environments, phosphates work better in acidic atmosphere. The bond is developed in a unshaped refractory by addition or in situ generation of phosphates, either by reaction of phosphoric acid (H_3PO_4) with metal oxides (say alumina) or by direct addition of phosphates [like, mono-aluminum phosphate, MAP or $Al(H_2PO_4)_3$] in the unshaped refractory composition.

When the first option is used, the phosphoric acid (H_3PO_4) initially reacts with Al_2O_3 (above 127°C upto 427°C) or with $Al(OH)_3$ (at room temperature) forming mono-aluminum phosphate (MAP) $Al(H_2PO_4)_3$. This MAP decomposes on heating to form ortho-aluminum phosphate (OAP) between 732°C and 1327°C, and which on further heating forms pure alumina and phosphoric pentoxide in gaseous state.

$$6H_3PO_4 + Al_2O_3 = 2Al(H_2PO_4)_3 + 3H_2O \text{ (Temperature 127°C–427°C)}$$

$$3H_3PO_4 + Al(OH)_3 = Al(H_2PO_4)_3 + 3H_2O \text{ (Room temperature)}$$

$$Al(H_2PO_4)_3 = AlPO_4 + 3H_2O + P_2O_5 \text{ (Temperature 732°C–1327°C)}$$

$$2AlPO_4 = Al_2O_3 + P_2O_5 \text{ (above 1350°C)}$$

So this bonding finally results in pure alumina with no trace of phosphorus in the fired composition. Hence, it produces no secondary phases. But there are several disadvantages associated with phosphates. Primarily the setting of the material is very slow and sluggish. In order to speed up the reaction, setting agents are used, like, MgO, CaO, calcium aluminates, etc. These additives induce an acid–base reaction, forming amorphous/crystalline phosphates that enhance the refractory hardening and other properties. Also, phosphate bonding requires considerable amount of mechanical moisture for workability, and so a long heat-up schedule is required for drying. Also the green strength is weak due to lack of strong bonding, like in cement. On firing, the bond system loses a considerable amount of strength at operating temperatures above 1300° till ceramic bond forms on sintering. This binder system has also poor ability to bond with existing fired refractories. In addition, high chance of lamination is there in the refractory lining, as they have weak bonding strength.

In the second type of bonding, dry phosphates are used with water. Most commonly available and used phosphate powder is MAP due to its high solubility in water, greater bonding strength, and reaction with basic and amphoteric raw materials at low temperatures. The performance of this bonding depends on the size of the phosphate powders, mixing process of dry phosphate with the refractory aggregate, and dissolution of phosphates in water and its reaction. This type of phosphate bond gives reasonably quick set times and moderate green strengths due to the reaction of the dry phosphate and water. Proper heat-up schedule is required to remove moisture, and to avoid cracking and spalling. Again, improper mixing and dissolution of phosphate will result in inconsistent bonding and properties of the unshaped refractory.

Phosphate bonding is used for alumina-based compositions but not used for silica or magnesia compositions. Silica does not react with phosphoric acid at low temperatures but forms $SiO_2.P_2O_5$ and $2\ SiO_2.P_2O_5$ compounds at high temperatures, which are low melting (1100°C–1300°C) ones. For magnesia, the phosphoric acid reacts instantly, and setting of phosphate is very fast. Also, different magnesium phosphates are formed, and all are low melting compounds, the maximum being newbeyrite ($MgHPO_4.3\ H_2O$) with melting point 1327°C. Hence, phosphate bonding is avoided in silica and basic systems.

13.3.3 SILICA FUME (FLOW MODIFIER)

The concept of unshaped refractory, especially the castables, has been changed with the reduction in cement content, which was only possible by using the flow modifier, silica fume. Use of silica fume has reduced the amount of cement from about 12–20 wt% level to 4–6 wt% level and improved the high-temperature properties without affecting the installation and low-temperature properties. Silica fume, also known as micro silica, is a by-product of silicon and ferro-silicon industries. It is a noncrystalline polymorph of silica having average size of particles below 0.15 micron, surface area ~20 m²/g, and spherical in shape. The main application area of silica fume is high-performance concrete, since about 1950s, for improvement of ability to flow due to its shape, size, and pozzolanic activity. As this silica fume consists of minimum 90% silica, which is a good refractory material, the use of silica fume in refractory was planned.

In unshaped refractory fume silica plays multiple beneficial roles, mentioned as below.

1. Microfiller: Fume silica is a submicronic superfine powder that can enter into the gaps and voids between the various sized aggregates and fines. This micro-filling activity enhances the compaction of the refractory, resulting in increased densification and strength. Also, this filling activity helps in the requirement of fine cement hydrates for filling and squeezes the water out from the cement grain, thus reducing the cement and water requirement of the unshaped refractory.

2. Ball-bearing effect: Morphology of the fume silica particles plays an important role in characteristics of green mixture of the unshaped refractory. Spherical-sized moist fume silica particles roll down in between the aggregate particles, creating a "ball-bearing" effect that decreases the friction between particles moving past each other more easily (Figure 13.3). This rolling action facilitates the movement of coarse particles, overcomes the interlocking action among coarse particles, and reduces the water demand of the refractory to a great extent. Silica fume improves the flowability and workability of cement. This effect reduces the unit water amount required for achieving the specific flowability.

3. Pozzolanic activity: Fume silica contains noncrystalline, submicron-sized silica particles that react with the $Ca(OH)_2$ phase produced by hydration of the cement. The resultant compound is calcium silicate hydrate (C–S–H). This hydrated gel is deposited in the open pores of unshaped refractory,

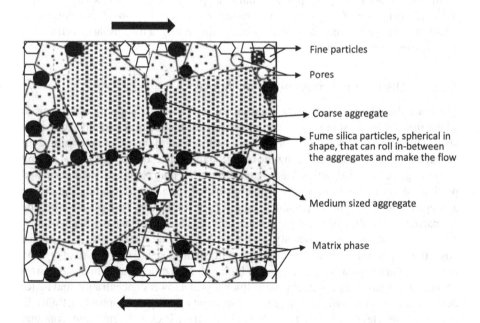

FIGURE 13.3 Ball-bearing effect of spherical fume silica particles to improve the flow.

reducing the porosity and thus improving the compaction, strength, durability, and corrosion behavior. This reaction also improves the water tightness of the cement, making a stronger monolithic.

Silica fume enhances the flowability of the unshaped refractory as similar to the concrete, mainly due to its spherical shape and finer size. Thus, it helps in reduction in water content, which is otherwise essentially required to get flow/movement in refractory mass. Also, finer sized silica fume particles easily enter in the voids, even they are very small in size and fill them, thus enhances the packing of the refractory, densification, sintering, and helps in strength development. They are also termed as microfiller. Hence, presence of silica fume reduces the requirement of the cement and its hydrate phases to fill the voids and packing, thus reduction in cement content is possible. Also, silica fume has some pozzolanic activity, reacts with water to form hydrated bond, mainly in the $CaO-SiO_2-H_2O$ and $CaO-SiO_2-Al_2O_3-H_2O$ systems. Also, during the formation process some negative charge remains on the surface of silica fumes, which helps to disperse the cement-containing systems, where cement particles are positive in charge. Thus, they also reduce the water requirement by satisfying the charge of the cement particles. Being a major source of silica in high alumina compositions silica fume also helps in to form mullite at high temperatures, which enhances hot strength, corrosion, and thermal shock resistances.

Typically a commercial silica fume contains silica in the range 90%–95%, but the major impurities present are carbon in the range of 0.2%–1.5% (coming from the silicon and ferro-silicon industry, used as raw material for carbothermal reduction), iron oxide in the range of up to 2%, alkali oxides (Na_2O and K_2O) up to 4%, alkaline earth oxide (CaO and MgO) up to 3%, alumina up to 1%, etc. Impurities are harmful in the performance of silica fume as a refractory material. Also, higher content of alkalis affects the setting behavior; it enhances the setting of the unshaped refractory, allowing less working time.

13.3.4 Dispersants and Anti-Setting Agents

These are the additives especially required for the cement-containing compositions where flowability is an important parameter, say for low-cement and advanced castables. Dispersants work on the conventional deflocculation theory of increasing double layer thickness, thus increasing the charge stabilization of the matrix system and increasing the flow properties. Dispersant containing matrix acts as a deflocculated particulate system that allows the whole unshaped refractory, including the coarse aggregates, to flow. The dispersants contain polar molecules that get absorbed by the particles, and they increase the surface charge. Thus, a surface repulsion occurs, resulting in a decrease in viscosity, and a stable, flowable matrix system develops. Also, the dispersants increase the pH range of the matrix stabilization. Use of organic polymer compounds as dispersants with high polymerization (large molecular size) also helps in increasing the double layer thickness at a lower concentration and better deflocculation. The common inorganic dispersants are pyrophosphates [$(P_2O_7)^{-4}$], tri-polyphosphate [$(P_3O_{10})^{-5}$], hexa-meta-phosphate [$(P_6O_{18})^{-6}$], etc., and common organic dispersants are citrates, poly-acrylates, pol-metha-acrylates, etc.

Anti-setting agents are those additives that delay the setting of cements; they are especially added to cement-containing compositions. Addition of water in a cement-containing unshaped refractory will instantaneously initiate the hydration of cement, and the cement will start setting. As the mixing of unshaped refractory with liquid is difficult at the exact location of application, a time gap between addition of water (liquid) and setting is required for transporting the mixed mass and installation of the material, which is technically called as working time. Instantaneous setting of cement will make the unshaped refractory hard, cause the minimum working time, and will result in a poor installation. Anti-setting agents are added to make the unshaped refractory workable for a sufficient duration even after water addition, and keep the mixed mass workable and flowable for proper installation.

Anti-setting agents are generally hygroscopic in nature; they absorb the moisture added to unshaped refractory and delay the cement hydration reactions. But, with time, cement particles draw out the moisture from these hygroscopic materials, and starts to hydrate and then set. This time delay is essential for better flow, placement and installation, and final properties of the refractory. But higher addition of anti-setting agent will delay the setting of cement for a longer time, resulting in poor or no setting, flow out or deformation of unshaped refractory, poor strength, and other properties. Hence, amount of anti-setting agent is very crucial for proper installation and property development of cement-containing unshaped refractories. The common anti-setting agents used are citric acid, oxalic acids, tartaric acids, etc., and their ammonium and sodium salts.

13.3.5 FIBERS

Unshaped refractories are added with different fibers for imparting some special characters required for final property development. They are as follows.

13.3.5.1 Organic Fiber

Addition of water in unshaped refractory during mixing has two primary functions. It initiates the hydration reaction of the binder, and it supports the flow or movement of the unshaped refractory composition. Free water used for flow and chemically bonded water of hydrated phases come out from the refractory body/shape during heating, both during drying and firing. As the heating is done from the outside, the surface of the shape gets heated/dried and after removal of moisture shrinkage is strong at the surface only. This results in reduced passage for the further moisture to come out from the interior through the surface of the shape. But at the drying temperature, the water is vaporised, a huge internal vapor pressure is created within the shape. This pressure, if higher than the dried/green strength of the product, will crack, break, or shatter the shape. This drying breakage is termed as explosive spalling during drying.

To avoid this problem, organic fibers are added to the refractory compositions that will melt and create a small passage for the removal of the water vapor during drying. Hence, these organic fibers are essentially needed to melt and evaporate out from the composition before the pressurized moisture vapor formation and need to be very thin in diameter. Hence, organic fibers of polyethylene (melting point

~120°C), polystyrene (melting point ~165°C), etc. are commonly used. These fibers have high aspect ratio, with dimensions ~6–10 mm long and ~10–30 µm diameter, and are used in the range of 0.02–0.1 wt%. Use of these fibers creates a channel in the unshaped refractory, helps the vapor to go out and nullifies explosive spalling, and allows higher heating rate for drying and firing of the unshaped refractory.

13.3.5.2 Metallic Fiber

These fibers are mostly chromium- and nickel-containing steel alloy, have high melting and deformation temperature, prepared by wire drawn, centrifuged out from the melt, milled from the block, etc. methods, used to impart some tensile character in the unshaped refractories. The fibers are about 20–30 mm long, 0.3–0.4 mm in diameter, and used in the range of 3–6 wt% in the unshaped refractory. These fibers are added to such compositions that are applied to thermal shock-prone areas. The fibers allow the refractory mass to remain in the structure even after cracking due to thermal shock by pinning effect. Thus, they can enhance the life and performance of the refractory. But the addition of metallic fibers affects the flow behavior, increases water demand, but improves strength and thermal shock resistance properties. Though these fibers have oxidation resistance till ~1200°C, as they remain within the unshaped refractory and not exactly exposed at the hot surface, they can be used for processes operated at much higher temperatures.

13.4   BRIEF DETAILS OF DIFFERENT UNSHAPED REFRACTORIES

13.4.1   CASTABLES

Castables are the most studied and commercially most widely used material among the different monolithic refractories. Starting with the development of calcium aluminate cement, refractory castables have progressed from simple mixes of different fractions of main constituents and binder to a complex and technical formulations, suitable for various critical applications with very specific and tailored properties. Shaped refractories are increasingly being replaced by castables in many applications due to the enhanced performance and ease of installation. In general, castables may be defined as a blended mixes of different fractions of main constituent with bonding agent and various additives, supplied in dried conditions as loose powders, and mixed with a liquid (usually water) at the user industry and vibrated, poured, pumped, or pneumatically shot into place to form the desired shape or structure that becomes rigid because of hydraulic bonding or chemical setting and then fired to complete the process.

Castables can be of different types depending on the

1. Main constituent (alumina, magnesia, silica, etc.),
2. Amount or percentage of main constituent (60% alumina, 70% alumina, 90% alumina, etc.),
3. Density (dense, insulating),
4. Bonding material (cement, sol, phosphate, etc.),

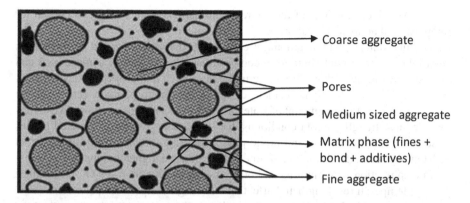

Coarse aggregate

Pores

Medium sized aggregate

Matrix phase (fines +
bond + additives)

Fine aggregate

FIGURE 13.4 Schematic microstructure of castable.

5. Amount of bond, especially for cement-containing one (conventional cement, low cement, no cement, etc.),
6. Flow and placement conditions (vibrating, self-flowing, etc.).

Castables are having, unlike shaped refractory, a continuous bonding or matrix phase (containing fines of the aggregate component, bond material and additives) wherein the aggregates are distributed (Figure 13.4). Hence, development of matrix phase is critical for the performance of the castables. So fines and additives play the major role in the property development and performance of the castables. Also, contrary to shaped refractories, castable need to flow during installation along with density and strength development. Again, density and strength development depends on the compaction among the particles resulting in reduced flow properties. So two parameters, flow and compaction, are contradictory in nature. Hence, conventional particle-packing models, namely discrete models, are not applicable for castable developments, and continuous particle size distribution models are used. This is important for the advanced high-performance castables where a reduction in water demand is a prime factor for performance. However, for conventional castables, discrete model with wide size distribution of aggregates are used to make the system and process simpler. Details of particles size distribution is discussed later.

Generally, a refractory castable contains about 40–80 wt% of aggregates and fines, 5–30 wt% of property modifiers, and 2–30 wt% of bond material and additives up to 5 wt%. The aggregates form the basic skeleton of the castable and account for the largest amount of the formulation. The sizes of the aggregates can range from 10 to 80 mm. Aggregates are sized and proportioned with medium and fine fractions to achieve the desired particle packing and distribution. Choice and chemical composition of the refractory aggregates and fines help to attain the desired chemistry and chemical properties of the castable, mineralogy, and also the physical properties. To improve the filling character to achieve the desired properties in castables, the fillers and modifiers play the most vital role. Multiple fillers are also used in many cases to attain proper distribution of particles and the final properties. Fillers and modifiers can be finer sized fractions of the aggregate minerals or other minerals chosen for compositional adjustment and property improvement.

Bonding material and its amount greatly affect the properties and performances of castables. For cement-containing compositions, the amount of cement used and its associate additives play a great role in the placement and properties development of the castable. Conventional cement castables commonly contain cement above 12 wt%, require a higher amount of water, and result in a porous structure. Also, they have greater strength in cold conditions due to a greater extent of hydrated phases present at unfired condition and liquid phase sintering at high temperatures. But at high temperatures, the formation of liquid phase due to a higher amount of CaO present in combination with alumina and other impurities like SiO_2, Fe_2O_3, etc., reduce the high-temperature strength and affect all other high-temperature properties. Also, decomposition of hydrated phases result in drastic deterioration of the strength at the intermediate temperatures (300°C–1000°C) till sintering occurs and a ceramic bond develops.

Low-cement castables, having cement in the range of 4–6 wt% with fume silica of similar amount, have improved the properties at high temperatures due to lesser amount of liquid phase present, but still formation of liquid phases are there, and castables are applicable up to 1550°C. Above this temperature, the formation of low melting compounds, namely, anorthite ($CaO.Al_2O_3.2SiO_2$) and gehlenite ($2CaO.Al_2O_3.SiO_2$) are high, and hot properties degrade drastically. Replacement of silica fume by reactive aluminas provides similar flow properties and improves the hot properties of the castables. Further reduction in cement to 1–1.5 wt% makes ultra-low cement castables, which further improved the properties by reducing the amount of liquid phase formation.

Cement-free compositions are developed to avoid the lime-containing liquid phases. Also, presence of lime in cement-containing compositions adversely affects the corrosion resistance. Use of silica sol makes castable with improved high-temperature properties, especially due to inherent mullite phase formation in high-alumina compositions. But, weak coagulation bonding of the silica sol results in poor strength at ambient temperatures. But, there is no degradation in strength at the intermediate temperatures for sol-containing compositions, and hot strength is considerably improved compared to any cement-bonded material. Hydrated alumina-bonded castables are special kind of materials where the whole composition, both aggregate and matrix part, can be made of alumina only (for high alumina compositions) and may result in excellent hot properties, including corrosion resistance. But weak bonding at low temperatures, huge amount of moisture loss (chances of cracking during drying), and poor sintering properties are the difficulties to attain the desired properties. Phosphate-bonded materials are not only excellent from workability point of view but also porous in character and have weak bonding strength at low temperatures.

The flow characteristic of the castables is important as a better flowing material can easily take the desired shape even with intricate design. But excessive high flowability may result in very low viscous material, which may cause separation of aggregate and matrix part. Poorly flowable mass is difficult to cast, shape resulting in poor placement, installation and final properties. External energy is supplied using vibration in most of the castables that allow the mixed composition to flow properly and take the intricacies of the shapes. Also, there are castable formulations that attain the desired flowability without any external vibration (flows under its own weight are called self-flow castable, where precise control of particle size and its distribution are very important). Flowability of the castable mix is important for the initial period (after mixing with water), and then it must decrease gradually with time as the material has to set (harden) and strength development occurs.

Among the properties, density and porosity values depend on how better the voids are filled and how well the composition is densified. In both the cases, fines and the matrix part play an important role. Thermal conductivity, abrasion resistance, and other properties of the castable increase with the increase in compaction. Few properties of castables are dependent on the constituents present. Presence of higher amount of alumina increases the density values and also imparts greater abrasion resistance. Also, higher room temperature strength will result in better abrasion resistance. However, at high temperatures, the abrasion resistance depends on the alumina content and the strength of the bond at those temperatures. Most of the high-temperature properties are dependent on the presence of liquid phase and densification or sintering of the composition.

13.4.1.1　Particle Size Distribution

Particle size distribution (PSD) is important for both shaped and unshaped refractories, but it is more critical for the castables as they need to satisfy two near-contradictory properties, flow, and compaction (strength). For conventional shaped refractories, discrete packing of particles is important, where, finer fraction of particles enter into the void space of the coarser portion and thus improve packing, compaction, and strength. But, for castable, the same is not appropriate as the mass needs to flow also. Flowability of the castables is very important for the development of final properties and their performances. It allows to form any intricate shape and a good lining. Again, flowability is primarily dependent on the PSD and packing of the castable constituents.

The basic unit for PSD is the particle size. A particle is a small piece or part of a material that represents the physical/chemical properties of the materials. Size represents an average dimension of a particle between its two extreme ends. Size is easy to define for any regular shape, like sphere or cube, where the diameter or side represents their size, respectively. But for an irregularly shaped particles, like for refractory materials, defining the average size is complicated. Table 13.3 describes the different ways for representing the size of an irregularly shaped particle equivalent to the diameter of a sphere of similar volume. However, commonly the diameter of equivalent sieve size is used to represent the particle size for refractories.

In castables, the aggregates are separated from one another by the matrix phase. This results in lesser friction between the coarse aggregate particles, also within the castable, making them to flow easily. Hence, the closest packing concept for the conventional shaped refractories does not work for castables. A new particle size distribution concept is used, called continuous particle size distribution. Here, the particles are present in a continuous manner from their size point of view, and each size fraction used for this type of distribution has a maximum and minimum size value. The maximum size of one particular size fraction is the minimum size of the

TABLE 13.3

Description of Size of an Irregularly Shaped Particle Correlated with the Diameter of an Equivalent-Sized Sphere

d_{max}	Diameter of the sphere same to the maximum dimension
d_{min}	Diameter of the sphere same to the minimum dimension
d_w	Diameter of the sphere having similar weight (of same density)
d_v	Diameter of the sphere having the same volume
d_s	Diameter of the sphere having the same surface area
d_{sieve}	Diameter of the sphere passing through the same sieve aperture as that of the particle
d_{sed}	Diameter of the sphere having the same sedimentation rate as that of the particle

earlier coarser fraction, and the minimum one is the maximum size of the next finer fraction. Thus, the whole range of particles is present within the castable in a continuous manner. Size fractions used are relatively large numbers and are closely sized screened fractions to fill out the continuous distribution pattern.

Originally, the continuous PSD has been developed for the construction (concrete) industries and results in a very good rheology/flow characteristic at relatively low water contents with good compaction, low shrinkage, and high strength values. Due to the similarity in processing conditions and property requirement of the castables, the same concepts are equally important and applicable for the castables. But, as the ceramic particles are not exactly similar in features to that of concrete industries, some assumptions are made for the ceramics particles. The common assumptions are as below.

1. Particles are dense, hard and rigid, incompressible and uncrushable during flow.
2. The shape and size distribution in each size range is uniform and without any segregation during flow.
3. The particles will obtain the minimum volume (maximum contact and compact condition) during flow without any internal friction.
4. There will be no reaction between no gas/volatile matter evolution and no inter particle attraction/repulsion effect.
5. No particle will change its volume or shape during flow or after shaping.

Furnas was the first person to propose a mathematical relationship and particle size distribution model for different types of grading, during 1920–1930, for better compaction with flowability for mortar and concrete industries. The continuous particle grading proposed was without any gap of particle sizes. In actuality, the model for continuous PSD, as proposed by Furnas, was only an extension of the discrete particles distribution model for the multicomponent system without any gap between the particle sizes.

Andreasen and Andersen proposed a simple model to calculate the continuous particle size distribution based on some empirical work done with the concept of similarity condition. For the similarity condition they assumed a granulation image of the particles and proposed that the granulation image surrounding the particles of different sizes must be similar in the whole distribution. As per their model, the particle array surrounding each and every particle present in the distribution, regardless of its size, is exactly similar. That means the particles in the distribution are having self-similar patterns, meaning the surrounding of each and every particle, regardless of its size, are the same from near as from far. The distribution will appear exactly similar in structure even on changing the magnification (zooming) of view for the distribution. As per Andreasen and Andersen model, for continuous particle size distribution, the particle size variations and packing arrangements will be exactly similar at any magnification in the distribution. Based on this similarity condition for continuous particle size distribution, they proposed a linear equation relation CPFT (Cumulative Percent Finer Than) plot against particle size.

$$CPFT/100 = (D/D_L)^q$$

where D = particle size of any fraction; D_L = the largest particle size in the castable composition / mix; and q = the distribution coefficient. They also concluded that the exponent or the distribution coefficient (q) in the equation plays a vital role in defining the particles sizes, and proposed the value of q to be between 0.33 and 0.5 to obtain the optimum packing density.

The main drawback of Andreasen and Andersen model was that they did not consider the effect of the smallest particle size in the distribution. As per their model, there will be distribution of particles even at infinitesimal small sizes, which is not feasible at any practical conditions (like subnanometer sizes). Hence, as attaining such small sizes is not feasible, this drawback restricts the model for accurate application and wide popularity.

Now, for any experimental work on PSD, the finest size (the minimum available and feasible size) must be fixed. So, packing is stopped at some finite minimum size, D_s, which means no finer sized particles are available in the distribution. Hence, for each real system, this D_s is defined. Dinger and Funk has introduced this smallest particle size (D_s) of the distribution for the calculation of particle size distribution, and modified the Andreasen and Andersen equation as follows,

$$CPFT/100 = (D^n - D_S{}^n)/(D_L{}^n - D_S{}^n)$$

Particle size distribution plot as per Dinger and Funk is identical to Andreasen and Andersen plot but has slightly higher percentage values (of each particle size class) due to normalization to 100% total volume after the smallest size, D_s, is defined and introduced. If we compare both the distribution model together in a same plot with same distribution coefficient (q value), Andreasen and Andersen plot will show higher CPFT values for any size in the distribution compared to that of Dinger and Funk. Meaning Andreassen and Andersen plot shows a finer distribution (Figure 13.5). According to the figure, if we take D_L as 1 mm (1000 μm) and D_S as 10 μm, then as per Andreasen and Andersen model there will be about 18.2% particles finer than 10 μm, whereas it is zero for Dinger and Funk model.

13.4.1.2 Classification and Properties
Refractory castables are classified in different modes. The primary classification is based on chemistry, the major constituent present, which differentiates the castables

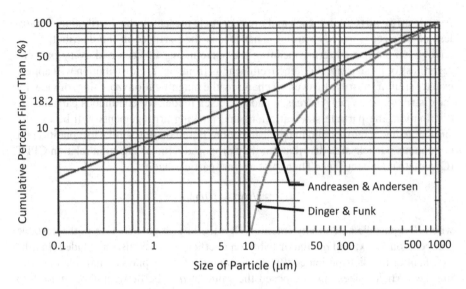

FIGURE 13.5   Comparison between Andreasen & Andersen and Dinger & Funk models.

TABLE 13.4
Classification of Castable as per Cement Content

Castable Classification	Cement Content	Amount of CaO
Conventional Castable	>12%	>2.5%
Low Cement Castable	Between 4% and 6%	Between 1.0% and 2.5%
Ultra-Low Cement Castable	Between 1% and 1.5%	Between 0.2% and 1.0%
No (Zero) Cement Castable		<0.2%

---

based on alumina, silica, magnesia, etc., content. Again, each category of the castables can be further classified as per the bond material used, as per the content of the
bond material, their flow characteristics, density, or insulating character, etc. The
compositions of each type of castable generally vary with respect to the content of
aggregates, fines, matrix components, bonding material, additives, etc., to attain the
desired properties. According to the type and content of the constituents, the properties of the castables also vary widely. However, like shaped refractories, the major
volume of castable used worldwide are based on alumina and alumino-silicate materials, and also some silica, basic castables, carbon-containing castables are used for
their certain specific properties and for specific applications.

According to ASTM C401–91, the classification of alumina and alumino-silicate-
based castables are primarily based on calcium aluminate cement content. Also, the
alumina content affects the properties grossly and makes different classes of castables. Table 13.4 shows the most common classification of castable based on cement
content.

A conventional castable is commonly based on alumina and alumino-silicate composition containing the highest amount of calcium aluminate cement. This type of castables is the most primitive one, and effect of particle size and other advancements are not considered. This type requires very high amount of water, usually 8%–20%, for flow and placement of the castable and also for the hydration of high amount of cement. This high amount of water results in huge porosity generation in the fired product, affecting mechanical, thermos-mechanical, chemical properties of the castables. Incorporation of particle size distribution concept in castables with the use of flow modifiers, like fume silica, has produced the low cement castables with reduced requirement of cement and water (between 4 and 8 wt%). The drawbacks of conventional castables were minimized. Further reduction in cement content, producing ultra-low-cement and no-cement castables, further improved the properties.

Constituents and chemistry directly control the property and performances of the castables. Tables 13.5–13.8 provides various types of catables that are commercially being used, and with their compositions and properties. From the tables it can be seen that for alumina and alumino-silicate castables with increasing alumina content the application temperature increases. Again, high in composition, especially iron and titanium oxides, they will deteriorate the high-temperature properties and the maximum service temperature. Also, with decreasing the cement (CaO) content, the application temperature also increases. Thorough study of the table will provide detailed idea about the effects of composition and constituents on the properties of different types of castables.

TABLE 13.5
Details of Different Commercially Used Conventional Cement Castables

Common Name	CC45	CC50	CC60	CC70	CC80	CC90	CC95
Al_2O_3	45	50	60	70	80	90	95
SiO_2	40	42	31	12	8	1.8	0.3
Fe_2O_3	3.5	1	1	4	1	1	0.3
TiO_2	5	1.2	1.5	6	1.5	1	
CaO	5.5	5	5	6	6	5	4
Water requirement, app. (%)	12	12	12	11	11	10	10
Bulk density at 110°C	2.12	2.15	2.25	2.55	2.8	2.85	2.9
Cold crushing strength (MPa) at							
110°C	40	40	40	40	60	60	65
1000°C	18	25	25	35	35	35	40
1350°C	30	–	–	50	–	–	–
1550°C	–	55	50	–	60	60	70
Permanent linear change at							
1300°C	−0.8	–	–	−0.2	–	–	–
1500°C	–	−0.8	−1.0	–	−0.8	−0.7	−0.8
Maximum application temp (°C)	1400	1500	1600	1450	1700	1750	1800

TABLE 13.6
Details of Different Commercially Used Low-Cement Castables

Common Name	LC45	LC60	LC70	LC80	LC90	LC95
Al_2O_3	45	60	70	80	90	95
SiO_2	48	35	25	12	5	3
Fe_2O_3	1	1	1	1.5	1	0.3
TiO_2	2	1.5	2	3	1	0.2
CaO	1.5	1.5	1.5	1.5	1.5	1.2
Water requirement, app. (%)	6	6	6	5	5	4.5
Bulk density at 110°C	2.3	2.65	2.75	2.9	3.02	3.05
Cold crushing strength (MPa) at						
110°C	70	75	75	80	90	95
1000°C	80	80	85	100	100	110
1550°C	100	100	110	120	120	120
Permanent linear change at						
1500°C	+0.4	−0.4	+0.5	+0.3	+0.2	+0.2
Maximum application temp (°C)	1500	1600	1650	1700	1750	1800

TABLE 13.7
Details of Different Commercially Used Insulating Castables

Common Name	A30	A40	A45	A90	Bubble A95
Al_2O_3	30	40	45	90	94
SiO_2	31	41	36	2	0.3
Fe_2O_3	6	4	4	0.3	0.1
TiO_2	4	4	4	0.2	
CaO	24	12	12	5	5
Water requirement, app. (%)	60	40	35	25	20
Bulk density at 110°C	0.8	1.2	1.5	1.65	1.6
Cold crushing strength (MPa) at					
110°C	1	4	12	7	10
1100°C	0.5	3	7	5	7
1300°C	0.8	6	9	7	10
1500°C					
Permanent linear change, 1100°C	−1	−0.3	−0.25	−0.1	0.2
Thermal conductivity, 800°C (kcal/h.m. °C),	0.16	0.3	0.4	1.0	0.8
Maximum application temp (°C)	1100	1300	1350	1750	1800

13.4.2   RAMMING MASS

Ramming mass is a type of unshaped refractory mainly used for installation and repair work, and are commonly a granular mass with semi-wet or -dry consistency. These refractories were very popular when the advanced castable was not developed. But with the development and advancement of LCC, ULCC and sol-bonded

TABLE 13.8

Details of Different Commercially Used Basic and Alumina–Spinel Castables

Common Name	Basic Castable		Alumina–Spinel Castable		
	Basic 80	Basic 95	A-Sp 80	A-Sp 95	A-M 95
Al_2O_3	10	1	82	92	92
SiO_2	3	0.3	9	0.8	0.8
Fe_2O_3	0.2	0.1	1	0.1	0.1
CaO	2.5	1	1.5	1.5	1.5
MgO	82	95	5	5	5
P_2O_5	1.2	–	–	–	–
Bond	Phosphate	Chemical	Hydraulic	Hydraulic	Hydraulic
Water requirement, approximately (%)	9	8	6	6	6
Bulk density at 110°C	2.7	2.8	2.95	3.05	3.0
Cold crushing strength (MPa) at					
110°C	50	50	60	60	60
1000°C	20	20	60	65	60
1600°C			75	100	80
Permanent linear change, 1500°C	−0.1	−0.1	−0.4	−0.4	+0.4
Maximum application temp (°C)	1700	1800	1700	1800	1800

compositions importance and use of ramming masses have decreased gradually. Ramming mass is mostly useful for cold applications and mainly used for the bottom part of the lining.

Most of the ramming mixes consist of refractory aggregates and semi-plastic bonding phase, which contains clay for the plastic property. The ramming mass gets kneaded when rammed under pressure, usually by pneumatic rammers. Various refractories aggregates, namely alumina, clay, magnesia, silica, zircon, etc., are used as per the chemical requirement of the application area. As ramming process involves some external pressure and the composition is semidry in nature and contains less moisture, the rammed lining exhibits lower porosity and better density. Nearly, no flowability or movement of the mass is required for installation. Hence, the particle sizes and their distribution for ramming masses are different than the castables. As flow is not important, less amount of liquid (water) is used for the installation. Also, the amount of fines is also less in the composition, whereas the amount of intermediate fractions is high. Also, the grain compaction is achieved during the ramming process by external pressure. The major advantages of ramming masse are as follows:

1. Better compaction
2. Lesser liquid (water), so shorter drying period
3. Faster installation
4. Installation can be done at a higher temperature than castable and others.

The binder system of the ramming mass is important, and it depends on the application conditions. Binders may be of organic and inorganic in nature. Organic binders used are cellulose, lingo-sulfonate, molasses, etc. Also, coal tar, pitch, resin, etc., are used as a binder for the systems containing nonoxide components, like carbon, as these binders contain a higher amount of fixed carbon and impart carbon in the composition. Organic binders produce lower strength after the burnout of the binder's carbonaceous matter. After burnout or oxidation, a relatively porous structure is obtained, resulting in reduced hot strength and poor resistance to corrosion, oxidation, abrasion, etc. Among various inorganic binders clay, phosphoric acid, phosphates, sol, boric acid, etc., are important. Sol-bonded compositions had lower initial strength due to poor strength of coagulation bonding, but strength increases with the increase in temperature. Also, sol bonding produces better high-temperature properties. Some details of commercially used ramming masses are provided in Table 13.9.

Ramming is always done by vertical build-ups of the material layer wise, so the compaction of one layer is achieved first and then the next layer is rammed. As the ramming process cannot impart great force, thickness above 100–150 mm cannot transmit the pressure to the material below, and so layer thickness is restricted. After each layer of rammed mass, the surface needs to be scratched to enhance the gripping or bonding between the different layers of materials and to avoid laminations. This is due to lesser self-bonding character of ramming masses. The amount of liquid to be used is also important as a dry mass will not compact well due to less bonding, and presence of excess liquid will make the material slide. In both the cases poorly dense porous structure will be produced. The rammed lining needs to be dried carefully

TABLE 13.9
Details of Different Commercially Used Ramming Masses

Common Name	MgO 80	MgO 95	SiO$_2$ 98	Al$_2$O$_3$ 90	A$_2$O$_3$ 95
Al$_2$O$_3$	10	1	0.5	90	98
SiO$_2$	3	1.5	98	8	0.2
Fe$_2$O$_3$	0.5	0.2	0.2	0.5	0.2
CaO	2.5	1	0.1	Trace	Trace
MgO	80	95	0.1	Trace	Trace
P$_2$O$_5$	1.2	–	–	–	–
Bond	Phosphate	Chemical	Ceramic	Chemical	Ceramic
Water requirement, app. (%)	9	8		5	4.5
Bulk density at 110°C	2.8	2.9		2.75	2.9
Cold crushing strength (MPa) at					
110°C	50	50		20	10
1000°C	20	30		30	15
1600°C				40	60
Permanent linear change, 1500°C	−1.0	−0.6		−0.5	−0.2
Maximum application temp (°C)	1700	1750	1700	1700	1800

and slowly so that the surface does not get hard and impervious to the moisture vapor coming from below layers; otherwise cracking will occur.

13.4.3 GUNNING MASS

Gunning mass or gunite mix is another class of monolithic refractory, similar in composition and formulation to castable but with finer particle sizes. Use of gunning mas is a well-proven repair technique of refractory that can be applied quickly and cheaply. It is used for the hot repair of ladles and melting furnaces as well as relining or cold repair of the back lining. It is a refractory mass with different size fractions of aggregates, bonding material, and additives, which is forced under pneumatic pressure and placed in the desired area, where the lining is to be done, using a special equipment having a mixer machine and a gun. There are two methods of applying a gunning mass, namely, dry gunning and wet gunning.

In the dry gunning process, the dry composition of refractory is partially moistened with a part of the total water used to reduce dusting and initiation of bond (cement) hydration. The dampened mix is then transported through a hose by pneumatic pressure. The hose is capable of withstanding relatively high temperature and is hung from the top of the kiln or furnace where the gunning lining is to be done. At the end of the hose, there is a nozzle assembly where water is mixed with the refractory stream, and the resulting mixture is gunned onto the desired area. Initial dampening is done by about 5% of moisture, and final mixing with water at the nozzle is done with about 5%–10% moisture. The position of the equipment may be adjusted with the help of a crane. A binocular may be used (for critical locations where a human cannot enter mainly due to high temperature) to identify the worn out (corroded) area where repair is required is used, and then the equipment is placed in the proper direction and position for the gunning. This method is highly dependent on the conveyance of the premixed mass.

In wet gunning technique, the refractory mass is moistened with the total required water in the mixer itself and then pumped through the hose using an eccentric screw or a piston pump. At the end of the hose, the material is dispersed with compressed air through the nozzle. For wet gunning, the machine or the hose may get clogged by the moistened refractory, particularly when the equipment is not in continuous use. Also, the equipment requires more intensive cleaning. With dry gunning system, the blockages in the conveying hose can be blown off easily by compressed air only.

The performance of the gunned lining is dependent on the refractory system used and the binder. The gunning (lining) thickness is typically maintained between 10 and 30 mm and can be used for operation after about 3–5 minutes of gunning. Standard gunning materials are of magnesia, alumina, or silica base. Three essential requirements for a good gunning repair are optimal moistening, homogeneous mixing with water, and a high-quality gunning machine that guarantees uniform conveying and gunning. Less moisture may result in higher dusting, poor bonding, fall of material, poor property development, and reduced service life of the gunning mass. The excessive moisture results in running off of the installed surface and poor performance. Also, higher water leads to clogging and hardening of the refractory

TABLE 13.10

Details of Different Commercially Used Gunning Masses

Common Name	MgO 80	MgO 85	Al_2O_3 50	Al_2O_3 70	Al_2O_3 90
Al_2O_3	1	1	50	70	90
SiO_2	6.5	8	40	22	4
Fe_2O_3	1.2	1.2	1.5	3	1.2
TiO_2	Trace	Trace	1.2	2	1
CaO	6	1.5	2.5	2.5	2.5
MgO	80	85	Trace	Trace	Trace
P_2O_5	1.5	–	–	–	–
Bond	Phosphate	Chemical	Chemical	Chemical	Chemical
Water requirement, app. (%)	12–15		10–12	10–12	10–12
Bulk density at 110°C	2.5	2.7	2.3	2.5	2.9
Cold crushing strength (MPa) at					
110°C	10	35	40	45	60
1000°C	25	30	35	30	50
1550°C				50	80
Permanent linear change, 1500°C	−5.5	−6	−1.5	−0.8	−1.2
Maximum application temp (°C)	1700	1700	1550	1700	1800

mass within the hose, especially for the quick binding systems. Hence, gunning is a skilled activity and is dependent on the skill of the installation technician.

Various bonding systems can be used for gunning materials. Alumina cement is useful for the low-temperature strength development. Lime content and the impurities present in the material affect the high-temperature properties and performance of the mass on application. Phosphate bonding shows better bonding of gunning mass with the metallic surfaces. Resin bonding is used for the nonoxide-containing systems, especially for carbon- and carbide-containing systems, like blast furnace trough. Colloidal silica is also a good binder for the gunning masses, and there is no degradation of strength at the intermediate temperatures. Sol bonding is also used for the blast furnace trough applications. Some details of the commercially used gunning mixes are provided in Table 13.10. Binder system may vary depending upon the installation temperature and final application temperature. For any bonding system, the target of gunning is to get a

1. Crack-free homogeneous lining,
2. Minimum dust generation and rebound loss,
3. Good adhesion of material with the development of high density and strength.

Gunning is a very common repair technique mostly used in the iron and steel industries. It allows hot repair of the lining, making the whole process economic with reduced downtime. Gunning repair is common for blast furnace wall lining, trough areas, steel converter, steel ladle, reheating furnaces, etc. Figure 13.6 shows the

A

B   C

FIGURE 13.6 (a) Gunning machine, (b) Gunning repair of reheating furnace, and (c) Gunning operation in a steel converter (BOF).

gunning machine (a), repair of reheating furnace (b), and hot repair of steel converter or basic oxygen furnace (c).

13.4.4 Plastic Mass

The first patent on monolithic refractory was on plastic/pliable refractory in 1914, so this material can be taken as first developed unshaped refractory. Plastic mass (also called as patching mass) is supplied in moldable, preformed blocks or slices as air tight condition and placed by hand pasting or ramming for a quick, economical, and emergency repair of the furnaces. Plastic mass can easily take any shape due to its high plasticity and generally without any shuttering former for placement and installation. Depending on the thickness of the walls, the slice/block or its part is broken off or cut out and placed (by pasting/ramming) into position. After installation, the plastic mass is leveled, and proper attention is required for the drying process, to avoid cracks and differential shrinkage, as the mass contains a higher amount of moisture or liquid. Also, hard crust surface formation must be avoided during drying to avoid cracking or spalling of the lining. These materials can be both air-setting and heat-setting types, depending on the application requirements.

Plastic masses are classified as per the following.

1. Aggregate type: The aggregates may be clay, high alumina, corundum, silica, magnesium, chromium, zircon, etc.
2. Binder type: It can be of different types – clay, sodium silicate, phosphates, organic binders
3. Setting character: Plastic can be divided as heat-setting and air-setting types.

Plastic mass is made up of two main components, (i) aggregate, like fireclay, bauxite, high alumina, mullite, etc., and (ii) a bond material that provides the plasticity and strength. Aggregate system is important for the final property development and decides the application area. Among bonding systems, the most common and oldest one is the plastic clay. It is widely used due to its wide availability and lower cost but is associated with high shrinkage due to moisture removal (dehydration of physically absorbed and chemically bonded water). Clay–sulfate bonds are also used due to its high strength and economy. Most of the plastic masses are phosphate bonded. Phosphate bond sets slowly keep the mass workable for a prolong time and allow easy placement. Organic bonds like resin, coal tar, pitch, molasses, etc., are also used but can leave carbon in the refractory mass, hence mostly useful for compositions containing carbon or nonoxide components. Details of some commercially used plastic masses are given in Table 13.11.

Heat (or thermo)-setting or heat-hardening plastic refractory mainly uses phosphoric acid or aluminum phosphate as the binding agent. As the use of binder in solution form may enhance reaction with the fine powders and accelerate the setting rate, certain setting inhibitors, preservatives, and other additives are to be added in the composition. The main types of additives include oxalic acid, tartaric acid, molasses, citric acid, lignosulfonate, etc. Thermos-setting plastics have high strength at ambient temperature and high temperature, good thermal shock resistance, and strong peel resistance.

Air-setting (or hardening) plastic masses generally use aluminum sulfate as binder. As aluminum sulfate is economically available compared to aluminum phosphate, the cost of air-setting plastics is relatively low. The binder solution can be directly added to the material and stirred evenly. The shelf life (storage time) of air-setting plastic masses is longer than 12 months. The main disadvantage of this type of plastics is that the strength is lower than that of heat-setting one both at room temperature and high temperatures. Also, the binder, aluminum sulfate, reacts with iron in the powder, producing hydrogen gas, causing the mud to bubble or expand, and drastically affecting the properties. To avoid such a condition, plastic refractory is not bogged.

The plasticity index of the plastic mass is usually maintained between 20% and 35%. If the index comes below 20%, it becomes hard and dry, and difficult to apply.

TABLE 13.11

Details of Different Commercially Used Plastic Masses

Common Name	A 50	A 60	A 70	A 80	Basic
Al_2O_3	50	60	70	80	3
SiO_2	42	35	20	12	7
Fe_2O_3	1.5	1.2	3	3	1
TiO_2	0.5	0.5	2	2	
CaO	0.5	0.5	0.5	0.5	2
MgO	Trace	Trace	Trace	Trace	75
P_2O_5	1.5	1.5	1.5	1.5	–
Cr_2O_3	–	–	–	–	10
Water requirement, app. (%)	12–15	10–12	10–12	9–10	7–10
Bulk density at 110°C	2.2	2.3	2.45	2.55	2.3
Cold crushing strength (MPa) at					
110°C	10	15	20	25	30
1000°C	30	30	35	40	50
1550°C	50	50	55	55	
Permanent linear change, 1500°C	−1.5	−1.2	−1.2	−0.8	−0.2
Maximum application temp (°C)	1600	1650	1650	1700	1750

Again, more that 35% plasticity index makes the mass too soft and difficult to bog, and the shrinkage on heating is also very high.

Property of the plastic mass varies with the quality of bond used. In general, the bonding strength of aggregate phase is greater than that of matrix phase containing the binder. The structural strength of plastic refractory is controlled by the matrix or by the bonding phase. Binding agent is the main component that determines the strength. Generally, some extent of deformable character of plastic refractory remains until it is completely hard fired. This helps the plastic refractory to retain have deformable character up to a high temperature and a higher work of fracture. Thus, a plastic refractory resists the propagation of crack and cracks growth in a better way. Also, it shows a better thermal shock resistance property.

13.4.5 SPRAY MASS AND SHOTCRETE

The spray mass composition and its application are very similar to gunning mass but with further reduced particle sizes. Generally, the maximum size used is 1 mm. This spray mass acts as a protective coating for the main lining and acts as the hot face refractory, withstanding all the abrasion, corrosion, and other effects. In spray technique, as the material is fine and spraying is done from a closer distance than gunning, the rebound loss and wastage of material is very minimum. Spraying can be done on a hot surface also.

The spray mass is applied by a specially designed spray machine in which water is mixed with dry composition and made into a free-flowing homogenous slurry. This fully wet slurry is then pumped to the spray nozzle through a hose where it is atomized with air and sprayed effectively on the surface where the lining is to be done. The total spray machine consists of,

1. A large hopper (capacity up to 3 ton of material),
2. A weighing system to precisely weigh the amount of material to mix,
3. A mixer machine (sometimes with automatic control) that can provide kneading action also with the provision of water addition (with automatic control) to develop desired flowability and consistency,
4. Pumping facility to push the mixed material with desired consistency through the hose pipe, and
5. A compressed air-operated spray nozzle.

The total amount of water used is about 15%–25% to impart flowing nature and spray ability in the refractory mix. After spraying the mass, heating of the lining is done on a predetermined schedule to remove the moisture first, and a slow schedule is maintained due to a higher amount of water present. Next, the temperature is increased gradually to 600°C–700°C for complete water removal, and the spray mass lining becomes dry and hard. This spay mass does not sinter well sintered with the main (permanent) lining on which it is applied due to its different particle characters and other properties. So a weak parting plane remains between the spay mass and the permanent lining that helps them to remove easily (deskulling), if required. Spray mass (MgO containing) is a common hot face lining for the tundish in steel industries, as shown in Figure 13.7a. Details of the common spray masses used in the industries are provided in Table 13.12.

The main advantages of spray mass are:

1. Only a few mm lining thickness is done to face the aggressive condition of the hot face.
2. Minimum rebound loss.
3. Less skull formation and easy deskulling.

A   B

FIGURE 13.7   (a) Basic spray mass-lined tundish and (b) Shotcreting repair is in progress.

TABLE 13.12

Details of Different Commercially Used Spray Masses

Common Name	M 65	M 75	M 85
Al_2O_3	0.5	0.2	1
SiO_2	25	15	8
Fe_2O_3	6	5	1
TiO_2	Trace	Trace	Trace
CaO	2.5	2.5	2.5
MgO	65	75	85
Bond	Chemical	Chemical	Chemical
Water requirement, app. (%)	18–22	18–20	18–20
Bulk density at 110°C	1.7	1.8	1.85
Cold crushing strength (MPa) at			
110°C	2	2	2.5
1000°C	3	5	5
1550°C			8
Permanent linear change, 1500°C	−2.5	−3.5	−2
Maximum application temperature (°C)	1600	1650	1700

4. Flexibility to use in cold/hot condition.
5. Less manpower required.
6. Can be applied either on shaped or unshaped permanent lining.

Very similar to spray mass is a shotcrete mass where an additional liquid bonding/hardening agent is mixed with the pumped moistened mass at the nozzle mouth. Both setting accelerator and coagulating additives can be used to set the shotcrete faster. This technique was originally developed for the construction industries and provides advantages like high installation rate, good mechanical properties, etc. Generally, a composition with self-flowing consistency is preferable for better pumpability and flow with low rebound loss. Shotcreteing as a repair mass on a permanent refractory lining is shown in Figure 13.7b.

Shotcrete is similar to gunning mass from technical and application points of view, but is applied with a further improved technology. In shotcrete, the composition is premixed with water, and a homogeneous mixture is obtained, which results in uniform properties all through the lining. Whereas, in gunning, water is mostly added at the nozzle, hence less time for homogenization that deteriorates the property of the lining. Again, due to lack of homogenization and mixing dust, generation is high with greater rebound loss for gunning materials. Hence, shotcrete produces good lining with the uniform property, without any lamination and voids structure.

13.4.6 Mortar

Mortars are used to fill refractory joints, bond refractory shapes together, and protect the joints from an attack of slag or other fluxes. They are used to complete the lining for bricks, preformed blocks or shapes, and also for insulating products. Mortars should be compatible with the main refractory that they are bonding and should not have excessive expansion or shrinkage compared to that lining material. Main functions of a mortar are:

1. Filling of the gap between shaped refractories,
2. Holding the individual shapes together,
3. Retain the definite shape and size of the total lining, and
4. To prevent corrosive and abrasive attacks.

The composition of these materials consists of fine aggregates and bond phase, which varies depending on the quality of the shape used and the field of application. Aggregate phase is similar to the shaped refractory material (sometimes the powdered form of the same shaped refractory) to match the properties with them. The binder may be of inorganic or organic in nature; among inorganic bonds clay, silicate, phosphates are used, and among organic bonds lignosulphonate, resin, tar, etc., are common. These materials are delivered to the user industries either in dry or a ready-to-use state. They are of two main types:

1. Heat-setting mortar, which hardens at elevated temperatures by chemical or ceramic bonds
2. Air-setting mortar, which hardens at ambient temperature by chemical or hydraulic bonds.

Mortars are relatively porous and weak in strength in comparison to the shaped refractories for which they are used. This is due to higher moisture content in them. This porous structure makes them flexible and helps to accommodate the volumetric changes of the shaped refractory on temperature fluctuations and heating cooling cycles. Some details of the commercially available refractory mortars are provided in Table 13.13.

TABLE 13.13
Details of Different Commercially Available Refractory Mortars

Common Name	A40	A 50	A 60	A 70	A 80	A90	M80	M90	MC1	MC2
Al_2O_3	40	50	60	70	80	90	3	1	3	5
SiO_2	52	40	32	22	12	4	10	2	2	2
Fe_2O_3	2	1.5	3	4	2.5	2		2	2	2
TiO_2	0.5	1	1.5	2	1.5	0.5				
CaO							2	1.5	2	2
MgO							80	90	65	40
Cr_2O_3									15	25
Service temp (°C)	1450	1500	1600	1650	1700	1750	1600	1700	1650	1600

13.4.7 DRY VIBRATABLE MASS

Dry vibratable mass (DVM) is a special type of unshaped refractory where no liquid or water is required for placement and installation. It is also a hot face refractory lining that acts as a protective layer for permanent lining and faces all the stringent conditions of the hot face, similar to spray mass. This was developed in the 1980s, close to the development of the spraying mass for the iron and steel industries, but did not become popular for many years due to its relatively higher cost and environmental issues.

The main advantages of DVM are:

1. Easy installation, as no mixing process is required with water.
2. Faster drying period, as no moisture removal.
3. Economic, as it saves fuel (no heat required to remove moisture).
4. No drying defects.
5. No hydrogen pickup (for molten steel during casting).
6. No direct adhesion with the permanent lining and easy deskulling.
7. Longer service life.
8. Free flowing in nature, helping the installation process and equipment simpler and easier.

DVM can be of both heat-setting and cold-setting types. For cold set system, dry powder is mixed with a liquid binder and hardener in a screw feeder before installation. Mixing of the additives is critical for the final performance of the lining. The heat-set materials are more flexible and user friendly. Dry material is directly poured in the gap area for lining between the permanent lining and former. The installation can be done by gravity filling or using a screw feed system. Gravity filling is a simple, low-cost, low-maintenance process but is a dusty process and prone to surging; whereas the screw feed system has a uniform filling process but costly with frequent maintenance requirement. During installation, vibration is done for better filling and compaction of DVM. After installation, the lining is cured with the former in place up to a temperature of 300°C. After heating, the material develops strength, and the former is removed after curing. The most common heating method practiced is the hot air blowing system using natural gas or LPG as the fuel. DVM lining also helps in reducing the energy cost. After the specific curing schedule, the former is removed, and the refractory lining is ready for operation.

The bond system plays the major role in installation and performance of these materials. Initial development was started with phenolic resin, but disadvantages like low installation temperature, use of a parting agent for smooth former separation, the evolution of harmful gases, cost, etc. are associated. Commercially available sodium silicate is also practiced, but demerits like low strength and the high sodium level potentially affect the performances. Abundantly available and cheap glucose ($C_6H_{12}O_6$) is being used widely due to its advantages like thermoplastic nature, installation even above 300°C (saving time and heat energy), and environmental friendliness. The most common application of DVM is in the tundish hot face/

FIGURE 13.8   DVM lining of tundish.

TABLE 13.14
Details of Different Commercially Available Dry Vibrating Mass

Common Name	M 65	M 75	M 85
Chemical analysis			
Al_2O_3	0.5	1.5	1.5
SiO_2	31	23	12
Fe_2O_3	0.6	1.5	1.8
CaO	6	5	3
MgO	60	70	80
Bond	Thermo-set	Thermo-set	Thermo-set
Bulk density at 300°C	1.75	1.8	1.85
Cold crushing strength (MPa), 1550°C	6	6	8
Application temperature (°C)	1650	1700	1700

working lining, as shown in Figure 13.8. Details of some commercially available dry vibratable mass used for tundish hot face lining are provided in Table 13.14.

13.5   MAIN APPLICATION AREAS

Unshaped refractories are becoming popular day by day in different types of furnaces and kilns, compared to the conventional shaped refractories due their multiple advantages. Use of these refractories helps in faster production (avoids the delay in manufacturing shaped one), simple and quicker installation, skill of brick laying is not required, etc. Monolithics are primarily important for the maintenance of furnaces and repair, saving greatly on time, investment and productivity loss. Certain monolithics also allow installation even under hot conditions, making the repair process friendly to production for user industry. Practically no preparation is required for installation of monolithics or repair a furnace by unshaped refractory. All these points, in combination, help unshaped refractories to replace the shaped ones very rapidly in different areas. A detailed information on application of different monolithics is given below.

Castables of different types and different compositions are important for various applications. Alumina-based conventional cement castables, containing alumina in the range of 45%–70%, are important for boiler, chimney, foundry ladle, sponge iron making, annealing, and heat treatment furnaces of different industries. Also, the high-alumina (>70%)-containing conventional castables are used for incinerators, abrading areas of sponge iron making, precalcinier of cement manufacturing, mouth of steel ladle, hot metal mixer, soaking pit, aluminum making furnaces, etc., low-cement castables with different alumina percentages (45%–90%) are important for sponge iron making and cement-making rotary kilns, tundish permanent lining, incinerators, aluminum-melting and -holding furnaces, lance pipes, laumders, snorkel of RH unit of steel plants, CFBC boiler lining, etc. More than 90% alumina-containing castables are important for fertilizer, chemical, petrochemical, carbon black industries. Very high alumina (>98%)-containing castables are used in carbon black reactor, coal gasifiers, and petrochemical industries. Alumina–spinel-based castables are used in steel ladle, seating and well blocks of ladles, porous plug, electric arc furnace delta and launder, tundish dams and weirs, etc. Alumina–silicon carbide–carbon-based castables, both with cement or colloidal silica bonding, are used in the blast furnace trough lining. Precast shapes of castable are used in ladle bottoms, burner block, dam and weir of the tundish, electric arc furnace spout and delta regions, well block, and setting a block of steel ladles, etc.

Insulating castable-containing alumina from 15 to 50 wt% are useful for petrochemical industries, especially as backup lining; heater, stack, the duct of petroleum reformer, etc.; and also in hydrocarbon process industries. High alumina-based plastic masses are useful for silver melting furnaces and acid regeneration plants. Alumina–silicon carbide–carbon-based plastic mass is useful for blast furnace tap hole. Alumina-based ramming mass and plastic mass are important for the repair work of EAF launders, petrochemical industries, hydrocarbon industries, etc. Low alumina (50%–60%)-containing plastic masses are important for incinerators and high-alumina (80%–90%)-containing plastic masses are used for aluminum-melting furnaces, acid regeneration units, silver melting furnaces, etc. Fireclay- and alumina-based mortars, containing alumina between 30% and 90%, are used for laying of shaped refractory. The alumina content of the mortar to be used should match with the alumina content of the shaped refractory being laid.

Basic castables (MgO content between 80% and 90%) are important for steel ladles and degassre units. Basic ramming mass (MgO cotent between 70% and 95%) are important for electric arc furnace cold repair, steel conveter tap hole repair, ladle back fill, protective coating for RH snorkel, induction furnace repair, etc. Basic gunning masses (MgO between 80% and 95%) are important for electric arc furnace slag line repair, steel converter hot face repair, RH degasser repair, etc. Basic spray masses with MgO content between 65% and 85% are used for hot face lining of the side walls and bottoms of steel tundish. Dry vibatable masses with MgO content between 65% and 75% are also used for the same application. Basic mortars with MgO content between 40% and 90% are used for lating of basic bricks, like magnesia, magnesia–chrome, chrome–magnesia, direct-bonded magnesia–chrome, etc. Dolomite-based ramming masses with MgO 35% and CaO 60% are used for backup filling of AOD vessel and ladle refining furnaces. Around 55% MgO and 40% CaO-containing dolomite ramming masses are used for cold ramming of AOD vessel, hot and cold repair of electric arc furnace hearth, etc. Dolomite-based gunning mass with 58%–60% MgO and 30%–32% CaO are used for repair of steel converters, steel ladles, electric arc furnaces, etc.

Silica-based ramming masses are important for the lining of the cupola of iron and steel industries, coke ovens, coreless induction furnaces. 98% silica-containing ramming masses are used for lining of coreless induction furnaces. 95% to 98% silica-containing mortars are used for laying of super duty and high heat duty silica bricks. Silica-based spray mass are used for hot repair of coke oven walls.

SUMMARY OF THE CHAPTER

Refractory materials that are supplied from the manufacturer's end without any shape are termed as unshaped or monolithic refractories. These loose, dry materials are usually mixed with liquid (commonly water) at the user site and then applied. Unshaped refractories are having many advantages over shaped refractories and replacing the shaped refractories in most of the industrial applications.

There are different ways of classifying unshaped refractories, and, among them, classification based on application (installation) technique is most widely accepted and used. These are castable, ramming mass, gunning mass, plastics, spray mass, etc.

Other than the conventional raw materials used for making shaped refractories, unshaped ones require some special materials. Property-modified fine fractions are important to control the matrix phase. Bonding materials are used to develop strength after shaping and also to enhance sintering. Property-enhancing additives are required, especially for the cement-containing compositions like silica fume, dispersants, anti-setting agents, different fibers, etc.

Various unshaped refractories are used in various industrial applications. Mostly, they are replacing the shaped refractories in all the different applications.

QUESTIONS AND ASSIGNMENTS

1. What is unshaped refractory? What are the advantages?
2. What are the different types of unshaped refractories used industrially?
3. Why we need bonding material in unshaped refractory?
4. What are the different bonding materials used for unshaped refractory?
5. What is the difference between building cement and refractory cement?
6. Describe the hydration mechanism of calcium aluminate cement.
7. Why developmental work continued to replace cement in unshaped refractory?

8. What are the advantages of silica sol as a refractory binder?
9. How does a silica sol help in the development of strength?
10. How does a phosphate bond work in the unshaped refractory?
11. What is a hydratable alumina? Detail the hydration mechanism of it.
12. Compare the different bonding systems used for unshaped refractories.
13. What is silica fume, and how it improves the properties of castables?
14. What are the functions of dispersant and anti-setting agent?
15. How does an organic fiber help? What is the role of metallic fiber?
16. Write in details on castables.
17. Compare ramming mass and gunning mass of unshaped refractory.
18. What is a plastic refractory? How does it work?
19. Write in detail about spray mass and dry vibratable mass.
20. Write about the different applications of unshaped refractories.

BIBLIOGRAPHY

1. ISO 1927: Refractory products: prepared unshaped dense and insulating materials classification. https://www.iso.org/standard/6625.html
2. C. Parr and Ch. Wohrmeyer, The advantages of calcium alumina cement as a castable bonding system, *Presented the St Louis Section Meeting of American Ceramic Society*, St. Louis, MO, 2006.
3. B. Nagai, Recent advances in castable refractories, *Taikabutsu Refractories*, 9 [1] 2–9 (1987).
4. T.A. Bier, N.E. Bunt and C. Parr, Calcium aluminate bonded castables: their advantages and applications, *Proceeding the 25th Annual Meeting of the Association of Latin-American Refractory Manufacturers (ALAFAR)*, Bariloche, Argentina, December 1–4, vol. I, pp. 75–84, 1996.
5. K.M. Parker and J.H. Sharp, Refractory calcium aluminate cement, *Transactions Journal British Ceramic Society*, 81 35–42 (1982).
6. W. E. Lee, W. Vieira, S. Zhang, K. G. Ahari, H. Sarpoolaky and C. Parr, Castable Refractory Concrete, *International Materials Reviews*, 46 [3] 145–167 (2001).
7. S. Banerjee, *Monolithic Refractories–A Comprehensive Handbook*, World Scientific/ The American Ceramic Society, Singapore, p. 311 (1998).
8. S. Banerjee, Recent developments in monolithic refractories, *American Ceramic Society Bulletin*, 77 [10] 59–63 (1998).
9. Y. Hongo, ρ-alumina bonded castable refractories, *Taikabutsu Overseas*, 9 [1] 35–38 (1988).
10. R. Racher, Improved workability of calcia free alumina binder alpha bond for non-cement castables, *Presented at the 9th Biennial Worldwide Congress on Refractories*, Orlando, FL, November 8–11, 2005.
11. W. Ma and P. W. Brown, Mechanisms of the reaction of hydratable aluminas, *Journal of the American Ceramic Society*, 82 [2] 453–56 (1999).
12. R. K. Iler, *The chemistry of Silica: Solubility, Polymerization, Colloid and Surface Properties and Biochemistry*, New York, Wiley, p. 866 (1979).
13. S. Banerjee, Versatility of gel bond castable / pumpable refractories, *Refractories Applications and News*, 6 [1] 1–3 (2001).
14. W.E. Lee, W. Vieira, S.Zhang, K. Ghanbari Ahari, H. Sarpoolaky and C. Parr, Castable refractory concretes, *International Materials Review*, 46 (3) 145–167 (2001).
15. J.E. Cassidy, Phosphate bonding then and now, *Bulletin of the American Ceramic Society*, 56 [7] 640–643 (1977).

16. R. Giskow, J. Lind and E. Schmidt, The variety of phosphates for refractory and technical applications by the example of aluminium phosphates, *Ceramic Forum International*, 81, E1–E5 (2004).

17. S..K. Das, R. Sarkar, P. Mondal and S. Mukherjee, No cement high alumina self flow castable, *American Ceramic Society Bulletin*, 82 [2] 55–59 (2003).

18. R. Sarkar, S. K. Das, P. K. Mandal, S. N. Mukherjee, S. Dasgupta and S. K. Das, Fibre reinforced no cement self flow high alumina castable – A study, *Transactions of the Indian Ceramic Society*, 62 [1] 1–4 (2003).

19. R. Sarkar, S. Mukherjee and A. Ghosh, Gel bonded Al_2O_3 – SiC – C based blast furnace trough castable, *American Ceramic Society Bulletin*, 85 [5] 9101–9105 (2006).

20. A. K. Singh and R.Sarkar, Effect of binders and distribution coefficient on the properties of high alumina castables, *Journal of the Australian Ceramics Society*, 50 [2] 93–98 (2014).

21. R. Sarkar, A. Kumar, S. P. Das, B. Prasad, Silica sol bonded high alumina castable: effect of reduced sol, *Refractories World Forum*, 7 [2] 83–87 (2015).

22. R. Sarkar and A. Parija, Effect of alumina fines on vibratable high alumina low cement castable, *Interceram*, 63 [3] 113–116 (2014).

23. R. Sarkar and A. Satpathy, High alumina self flow castable with different binders, *Refractories World Forum*, 4 [4] 98–102 (2012).

24. R. Sarkar and A. Mishra, High alumina self flow castable with different cement binders, *Refractories Manual*, 107–111 (2012).

25. R. Sarkar and A. Parija, Effect of alumina fines on high alumina self-flow low cement castables, *Refractories World Forum*, 6 [1] 73–77 (2014).

26. R. Sarkar and S. K. Das, Effect of distribution coefficient on gel bonded high alumina castable, *IRMA Journal*, 43 [1] 31–36 (2010).

27. R. Sarkar, Particle size distribution for refractory castables: a review, *Interceram: International Ceramic Review*, 65 [3] 82–86 (2016).

28. C. C. Furnas, Grading Aggregates, *Industrial and Engineering Chemistry*, 23 [7] 1052–1058 (1931).

29. A. H. M. Andreasen and J. Andersen, Ueber die Beziehung zwischen Kornabstufung und Zwischenraum in Produkten aus losen Körnern (mit einigen Experimenten), *Kolloid-Zeitschrift*, 50 217–228 (1930).

30. D. R. Dinger and J. E. Funk, Particle packing I – Fundamental of particle packing monodisperse spheres, *Interceram*, 41 [2] 10–14 (1992).

31. D. R. Dinger and J. E. Funk, Particle packing II – Review of packing of polydisperse particle systems, *Interceram*, 41 [2] 95–97 (1992).

32. D. R. Dinger and J. E. Funk, Particle packing III – Discrete vs continuous particle sizes, *Interceram*, 41 [5] 332–335 (1992).

33. D. R. Dinger and J. E. Funk, Particle packing IV – Computer modeling of particle packing phenomena, *Interceram*, 42 [3] 150–152 (1993).

34. J. B. Johnson and K. J. Saylor, A comparison of disposable magnesite tundish lining systems, *Presented at the 73rd ISS steel making conference*, Detroit, MI, 1990.

35. M. J. Bradley, Overview of tundish dry vibe technology, *Metallurgical Plant and Technology International*, 2 70–72 (2008).

36. M. W. Vance and K. J. Moody, *Steel Plant Refractories Containing Alpha Bond Hydratable Alumina Binders*, Alcoa Technical Bulletin, Alcoa Industrial Chemicals, Pittsburgh, PA, 1996.

37. P. Tassot, Cold setting DVM for tundish a real alternative, *Proceedings of the 3rd International Conference on Refractories at Jamshedpur, India (ICRJ)*, Jamshedpur, p. 187–191 (2013).

38. A. K. Singh and R. Sarkar, High alumina castables: a comparison among various sol-gel bonding systems, *Journal of the Australian Ceramic Society*, 53 [2] 553–567 (2017).

39. A. K. Singh and R. Sarkar, Urea based sols as binder for nano-oxide bonded high alumina refractory castables, *Journal of Alloys and Compounds*, 758 140–147 (2018).

40. R. Sarkar and A. Dhall Samant, Study on the effect of deflocculant variation in high alumina low cement castable, *Interceram- Refractories Manual*, 65 28–34 (2016).

41. R. Sarkar and J. Srinivas, Effect of cement and sol combined binders on high alumina refractory castables, *Refractories World Forum*, 8 [4] 73–78 (2016).

42. R. Sarkar, Nanotechnology in Refractory Castable: an overview, *InterCeram: International Ceramic Review*, 67 (S1) 22–31 (2018).

43. R. Sarkar and U. K. Behera, A study on the phosphate bonded high alumina castable: effect of MgO addition, *InterCeram: International Ceramic Review*, 67 (S1) 44–49 (2018).

44. R. Sarkar, Binders for refractory castables: an overview, *InterCeram: International Ceramic Review*, 69 [4–5] 44–53 (2020).

14 Trend of Refractories and Other Issues

14.1 PROGRESS IN REFRACTORIES

In the constantly advancing world of manufacturing, to say that refractories play an important role would be an understatement. In fact, refractories are one of the most essential components for all high-temperature processing. No one can even think of operating any high-temperature process without the support of proper refractories. Refractories influence significantly on the processing technology, product quality, and cost involved for any high-temperature operation. Refractory technology is constantly improving and upgrading itself as per the diversifying demands from the user industries, especially for the better product quality and productivity. To meet the demands of the user industries, refractories have also improved in qualities to endure harsh operating conditions of the user industries. With time, refractories are upgraded for prolonging the lifetime of furnaces and kilns and to meet the needs for increased productivity along with overall cost reduction. The refractory technology has also got advanced in repairing the damaged walls/linings of the old furnaces or ovens, even without affecting the production/operation much. To meet the demands of the user industries, the required quality parameters for refractories are becoming much stringent and higher than before. Also, all the high-temperature manufacturing processes have advanced in terms of technology, and the user industries are looking for higher production from the same furnace and with prolonged life. And the refractories are responding positively to all such demands.

Refractory technologists are working continuously to meet all the challenges and requirements of the user industries, and continuously upgrading the refractories in terms of quality, technology, concepts, newer products, raw materials optimization, waste utilization, and so on, keeping the environmental issue in mind. The improvements in refractories have played an important role in total demand and consumption of refractories per unit scale of product manufactured. Thus there is a continuous decrease in specific refractory consumption per ton (or unit scale) of production, resulting in continuous increase in furnace performance, productivity, and economy of high-temperature operations. This consumption rate has gone down continuously in all the user industries and has come to a bare minimum value. Table 14.1 lists out the global average for specific consumption of refractories in different industries. Also, improvement in quality of the refractories have led to much higher operating temperatures, producing high-quality products with greater productivity.

The specific consumption rate for refractories for all the different industries was enormously higher even few decades ago. Constant improvement in refractory quality with proper coping up with the technological upgradation in the user industries have made the developmental works on refractories successful and resulted in

DOI: 10.1201/9781003227854-14

TABLE 14.1

Global Average for Specific Consumption of Refractories in Different Industries

Industry	Operating Temperature (°C), maximum	Specific Consumption (kg/ton)
Cement	1550	0.8–1.2
Iron and steel	1700	10–15
Glass	1650	4–6
Aluminum	1250	5–8
Copper	1350	3–5

significant reduction in consumption rate. As a case study, let us take the refractory consumption scenario for iron and steel industries in Japan. Data and literature says the specific consumption of refractories was about 130 kg per ton of steel produced in 1950s, which was reduced to around 80 kg in 1960s, 30 kg in 1970s, below 20 kg in 1980, and currently it is 7–8 kg in 2020. This drastic reduction in consumption rate is not only limited to Japan and to iron and steel industries, but it is also a global trend for all the high-temperature processing industries. Only the specific values, timeline, and progress in technology are different. Such dramatic reduction in consumption rate of refractories for iron and steel industries, the case being mentioned here, is possible mainly due to the following reasons.

1. Upgradation in process technology for iron and steel production, like, conversion from open-hearth furnaces to basic oxygen furnaces, ingot making to continuous casting, etc.,
2. Introduction of different highly functional refractories for specific requirements, like porous plugs for steel cleaning, slide gate refractory to control the steel flow, etc.,
3. Conceptual change and upgradation in refractory technology, like, greater use of basic refractories, use of purer, even, synthetic raw materials, introduction and use of carbon and other nonoxide components in refractories, etc.,
4. Use of completely new refractories like, MgO–C refractory, zirconia-containing refractories, etc.
5. Development, application, and upgradation in unshaped refractories,
6. Greater use and significant improvement in refractory repair technologies.

The upgradation activities are continuous in nature for both the user and refractory industries, and the refractory industries need to continuously put attention for the betterment of their own and also for the user industries. Further, progress in refractory technology for all the high-temperature processing industries is going on in terms of new concepts of refractory lining targeted to reduce the lining thickness, resulting in increase in furnace volume, reduction in thermal mass of the furnaces, and increase in productivity and thermal efficiency. In the current scenario, developmental activities are also at commercial level to reduce the carbon (graphite) content in carbon-containing refractories to reduce the heat loss, reduction in chances of carbon pick-up by processed steel, and to take care of the environmental issues. Newer combination of refractories with the use of various additives, especially nonoxides for specific targeted applications, are in the forefront of the refractory research activities. New design of refractories, use of indigenous and recycled materials, greater extent of use of unfired and monolithic refractories, etc. are some of the prominent areas of current refractory developmental activities.

14.2   NANOTECHNOLOGY IN REFRACTORIES

Progress in refractories is a continuous process, and the prime impetus for the improvement in refractory quality is to perform better for a longer period. This target has also resulted in the development of new high-temperature refractory technology and products that perform better even under severe operating conditions, like, temperature, chemical environment, wear, and erosion, etc. Further, the refractory user industries have expectations of continuously improving the service properties along with reducing the cost of production and also maintaining ecofriendliness, waste utilization, scope for recycling, etc.

For the last few decades, significant success has been achieved by the refractory technologists in optimizing and adjusting the composition of suitable refractories to get superior physicochemical properties for specific operating conditions. Further control on properties and performance require greater control on the microstructural developments. Uniformity in microstructures with proper distribution of the constituents, even in nanometric scale, is very important to obtain the best results out of the composition. To have a greater control on the development of ideal microstructure, use of nanomaterials has started since long. Carbon in the form of soot, silica in the form of microsilica, aerosol, etc. are in use in refractories to improve various properties. Development of newer refractories containing various nanomaterials also helps to grow the market for nanomaterials and development of new industrial nanomaterials, stimulating an increase in the volume of their production and reduction of cost.

Global interest on nanomaterials has grown due to the significant change in properties of normal substances by converting them to a nanosize condition. Reduction in size in turn produces significant increase in surface area that results in increase in the proportion of atoms or molecules present on the surface. This again leads to a significant increase in the surface energy of the nanomaterial-containing material system, resulting in greater and faster physico-chemical processes and reactions with the surroundings.

Nanotechnology has significant importance in the refractory field, as the use of nanoparticles reduces the microstructural scale to the nanometer range with an enhancement in properties. Improvement in properties for the nanoparticles containing refractories is also due to the size of the defect that grows during failure, which becomes smaller due to the presence of nanoparticles compared to the size of the components. Thus causing much reduced damage to the refractory system. Again, the presence of nanoparticles will reduce the sintering temperatures of the refractories due to high surface energy and increase in chemical reactivity. Increased chemical reactivity also increases the possibilities of new compound or product formation, which is otherwise difficult or impossible in traditional system from the thermodynamic point of view. Nanoparticles present in refractory composition fill the gaps among the different-sized aggregate particles, enhance the compaction, reduce the porosity, and generate faster diffusion paths for sintering. Again, for refractories sintering is very important for the properties' development and performance. The matrix phase plays the most vital role for sintering and so the constituents of the matrix phase, namely, fines and additives, are most important for the property development and performance. For proper and uniform development of the matrix phase and to control the desired properties developed in refractories, nanoparticles are found to have great influence, as they have a significant control on sintering. Use of nanoparticles leads to enhanced densification, strength (both at ambient and elevated temperatures), resistances against wear, abrasion, corrosion, thermal shock, oxidation (for nonoxide-containing compositions), etc., for the refractories resulting in enhanced performance and service life.

If we look into the literature, we will find that the modern use of nanoparticles in refractories were started around the 1990s in Japan, especially to enhance the rheological and flow properties of monolithics, their strength and wear properties, and to reduce porosities. Similar advantages were also obtained for MgO–C refractories used in steel converters, where finer (nanosized) metal antioxidants increased the reaction rate for carbide formation, which not only prevents the oxidation of carbon but also improves the hot strength. Similarly, the use of nanocarbon in place of graphite has also significantly reduced the total carbon requirement for MgO–C refractory. This nanocarbon incorporation produces a reduction in the modulus of elasticity, improved thermal shock resistance, excellent corrosion resistance, increased hot strength and good oxidation resistance in the MgO–C refractory at a much lower (one-third to one-fourth) total carbon content. Reduction in carbon content also reduces the thermal conductivity of the refractory, preventing heat loss and reducing the chances of carbon pick-up by the processed steel. Again, the use of highly reactive nanoparticles opens up a vast range of possibilities as sintering agents in nearly all types of refractory products.

In the early days, use of nanoparticles were considered to be a costly affair; however, with the progress of time several advanced technologies have come up, producing nanopowders at much reduced cost, and use of nanopowders in refractories is commercially feasible. A variety of nanomaterials are already used or have the potential to be used in refractory field. Important among them are nanocarbon black (high, pure elemental carbon in nanometric particle size having semiamorphous molecular structure presently being used in tyre industries), carbon nanotubes (single-atom-thick graphite layers forming a long seamless cylinder with nanometric diameters), nanosized metal powders (as antioxidants), colloidal silica, etc. Commercialized technologies are also available to produce any type of oxide in nanometric size range useful for refractories, like, alumina, magnesia, zirconia, chromia, mullite, spinel, etc. whose addition accelerates the sintering behavior of the refractories, and also improve the properties and performances. Use of carbon nanotubes in submerged refractories has increased the cost by about 5% and resulted in an increased resistance against wear, erosion, and thermal shock resistance by about 15%–20%. Addition of oxide nanopowders in the range of 0.05–0.2 wt% in low-cement–high-alumina castable, especially used for iron and steel industries, has resulted in an improvement of strength by 25%–30%. However, even in the current days, the problems associated with these nanomaterials are their cost, availability, handling, mixing and uniform dispersion, and health and safety issues.

Use of nanomaterials in various fields are being tried, including refractories, for multiple benefits. But, it is obvious that every problem cannot be solved by the use of nanoparticles in the system. In refractories, use of nanoparticles and their proper and uniform dispersion in the system gives a chance in manipulating the properties even at the near-atomic level and thus opens up broad perspectives for improving their properties. This will also help in resolving various difficulties associated with refractories and their user industries, thus providing chances for new, innovative, highly competitive refractories' development.

14.3 ENVIRONMENTAL ASPECTS

Refractories are essential to make the furnaces and kilns, and to make the refractories, various raw materials are required, consuming the natural mineral resources of the earth. The mineral source of all the different refractory raw materials is the crust of the earth, and the mining of these raw materials increases the chances of mixing of these with the cultivation land, ground water, environment, and air, etc., causing serious threat to the human settlements close to the mining and mineral-processing industry. Again, traditionally, as the refractories are fired bodies, they consume huge amount of heat for manufacturing. The energy load of furnaces used in the manufacturing of refractories is quite large. Most of the environmental impact of refractories manufacturing is due to the generation of CO_x gases (primarily CO_2) produced from this high energy load. To improve this situation, developmental activities are focused on furnace systems, and techniques to control and reduce the energy load.

The refractory manufacturing process consists of four major stages, namely, powder preparation or mixing, shaping, drying, and sintering or firing. Energy audit for different types of refractory manufacturing process showed that about 65%–70% of the total energy required to make a refractory is consumed only in the firing process. Various types of refractories are fired at different temperatures, and the energy requirement for making different refractories are different. Table 14.2 provides the amount of energy required (in terms of average consumption of coal equivalent, in kg) per ton of fired refractory produced. The heat efficiency of the firing system is being improved by using waste heat recovery system, application of light-weight refractory, insulating refractory, emissive coating on the hot face (improving thermal radiation heat transfer efficiency), etc. Any such improvement in thermal efficiency will directly reduce the fuel consumption and also reduce the environmental impact by producing lesser amount of CO_2. Also, to minimize the serious threat to environment, multiple developmental activities are going on to use alternative sources of energy to manufacture the refractories.

In comparison with the traditional shaped refractories, monolithic refractories have been tremendously developed in the last few decades all over the world due to multiple advantages, including much reduced energy requirement. Significant reduction in energy requirement for monolithics is primarily associated with no shaping

TABLE 14.2

Energy Required for Production of per Ton of Refractory

Refractory Product	Energy (Coal Equivalent, kg of Coal) Required per Ton of Refractory Making
Silica bricks	338
Fireclay bricks	172
High alumina bricks	318
Low-grade magnesia brick	184
Medium-grade magnesia brick	232
High-purity magnesia brick	256
Magnesia–chrome brick	205
Direct-bonded magnesia–chrome brick	268
Magnesia–spinel brick	245
Magnesia–carbon or magnesia–alumina–carbon brick	31
Monolithics	10

and no firing process involved in the manufacturing. Also, in comparison to shaped refractories, monolithics have simple manufacturing process, shorter production cycle, easy installation technique, reduced material consumption by partial repair and relining on residual lining, suitable for making complex shapes, and convenient in adjusting composition and properties as per the requirements of the installation and application site. Multiple benefits by using of monolithics over shaped refractories, especially significantly reduced energy requirement and impact on environment, have enhanced its use to a great extent over the shaped ones. Japan was the first country to use greater extent of monolithics over shaped ones in 1992 and crossed 70% of monolithics use in total refractory consumption in 2011. The USA and many of the European countries are also using more than 50% of monolithics in total refractory consumption.

From energy consumption point of view, as mentioned in Table 14.2, monolithics require very minimal amount of energy to manufacture, even for precast shapes that require low-temperature heat-treatment process. So, from the energy requirement point of view, the monolithics can be termed as "green refractories." To generate heat, we need to burn carbonaceous fuel, and 1 kg of carbon on burning produces about 3.8 kg of CO_2. Again, for standard coal, it is about 2.5 kg of CO_2 per kg of coal (depending on fixed carbon content of the coal). So, on an average, manufacturing of monolithics saves more than 200 kg of coal per ton of monolithics (Table 14.2), which, in turn, saves the environment from more than 500 kg or more of CO_2 generation.

To keep the environment clean and safer for longer periods, attention have been given by the refractory manufacturers since long. Among multiple actions, few are described as below.

14.3.1 ENVIRONMENT-FRIENDLY (HAZARD-FREE) MANUFACTURING PROCESS

Refractory manufacturing process is involved with multiple stages that generate lots of fines and noise causing ecological and environmental pollution. Operations like, crushing, grinding, milling, blending, mixing, etc., are harmful for environment and manufacturers of many of the countries, and their workforce do not have sufficient awareness and measures for dust clearance and protection. For example, in silica refractory production, the workers are exposed to dreaded silica dust and are likely vulnerable to silicosis, if proper protection and care are not taken. Enough attention and care are must for the dust-generating units, and reduction or elimination in dust-related hazards can be done by wet processing, dust generator isolation, use of personal protection/safety equipment, protocols, etc. Separation of dust and fines from the flue gas is also important, as these hazardous fines can spread over a wide region, and can cause different environmental and health related issues. Bag (house) filter, electrostatic precipitator, ceramic filter, etc., are essential to prevent the flying off of the dust and fines into the environment. Also, use of waste heat recovery system, like regenerator, recuperator, etc., is important to save the environment from getting up heated by the hot flue gas, and it also helps to increase the thermal efficiency to a great extent.

14.3.2   USE OF ECOFRIENDLY (HAZARD-FREE) RAW MATERIALS

Raw materials are the basic constituents of refractories, and they control the properties and performance. To produce an environment-friendly refractory, it is essential to use all the raw materials, additives, and the manufacturing process ecofriendly. For example, in case of impregnated refractories, tar and pitch impregnation is to be done to reduce the surface porosity and to provide a carbon coating on the surface. But gaseous products generated during the processing, and fumes generated during heat treatment contain polycyclic aromatic hydrocarbons (PAH) like, benzo-alpha pyrenes (BAP), which are harmful to health and carcinogenic (cancer causing). So the processing of such refractories are to be done in isolated locations. Also, use of binders like sulfate and chlorides are also harmful, as they release toxic gases on heating. Hence, it is important for both the refractory producers and users to use raw materials and processing that are environment friendly and nonhazardous to human health.

14.3.3   CHROMIUM-FREE REFRACTORIES

Cr_2O_3-containing refractories are important for different industries, namely, steel, cement, glass, and nonferrous industries, mainly due to their excellent thermal shock resistance and superior corrosion resistance against molten slag. Chromium-containing refractories have much lower contact angle, especially against the silicate slags, resulting in reduced slag contact area and improved corrosion resistance. Due the superior properties obtained, Cr_2O_3-containing compositions were preferred for better performance, longer life, and higher productivity in user industries. However, during use, the transition element chromium under reduction–oxidation condition at high temperatures and temperature fluctuations may convert to Cr^{+6} species from its stable Cr^{+3} state. This conversion is again accelerated in the presence of alkali, lime in the atmosphere at high temperature. The Cr^{+6} state, known for its carcinogenic character, is water soluble, and contaminates the environment and exposes the humans and animals vulnerable to cancer. Chromium-containing used refractory commonly contains 0.1%–0.5% (1000–5000 ppm) of Cr^{+6}, and for any safe use of water, Cr^{+6} contamination must be below 0.05 mg/L.

Due to the above reasons, though technically Cr_2O_3 produces excellent properties in refractories, globally its use has been reduced since long. Many of the advanced countries, like the USA, many European countries have strictly restricted or banned its manufacturing and use. Chrome-free environment and ecofriendly refractories are being developed for every single application and are getting popular. Like, for cement industries magnesia–chrome refractories are replaced by magnesia–spinel bricks, magnesia–hercynite refractories, dolomite zirconia compositions, MgO–pleonastic spinel (containing Fe_2O_3–Al_2O_3 system) refractories; steel ladles are lined with alumina–spinel and alumina–magnesia castables; for RH snorkel, alumina–magnesia castable can replace MgO–Cr_2O_3 brick; for secondary steel refining MgO–Al_2O_3–TiO_2 and MgO–ZrO_2 refractories can replace MgO–Cr_2O_3 bricks, etc. However, this awareness and governmental restrictions are less in developing and underdeveloped countries, and proper attention and complete replacement of the chromium-containing refractories are highly time demanding.

14.3.4 ENVIRONMENT-FRIENDLY CERAMIC FIBERS

Ceramic fibers are a group of manmade vitreous (mostly silicate) fibers, used as insulation material for kilns and furnaces of different high-temperature applications, like, ceramics, iron and steel, nonferrous metals, metal treatment and foundries, petrochemical, chemical, and automotive industries. They are also used as insulator on the outside of the ceramic molds in investment casting to regulate the rate of cooling of the metal after casting. Commonly, these fibers are based on alumina–silica compositions and have the problem of nondegrading character when inhaled by human. The fibers irritate the skin, eyes, and upper respiratory tract and are fine enough to penetrate deep into the lungs and may lead to the development of lungs cancer. As the fibers are harmful, their uses are restricted in many of the developed countries.

Hence, high-quality heat-insulation materials are required to replace the common ceramic fibers, and developmental work is going on to prepare light-weight castables all through the globe. Calcium hexa–aluminate-based (not harmful to humans) light-weight castables also provide excellent thermal insulation character at high temperatures, even better than ceramic fibers. Trials are also conducted for this castable in applications like, ladle cover, reheating furnaces, etc., with encouraging results. Also, biosoluble ceramic fibers are in developmental stage that can be used as conventional alumina–silicate ceramic fiber persistently at elevated temperatures and that can get dissolved in pleural fluid. Once dissolved, the fibers cannot reach the lungs and cause any harmful effect, even if the same is inhaled by human.

14.4 RECYCLING OF REFRACTORIES

The recycling of refractories has been practiced by the industry for about half a century, but was not documented and recorded properly, as it was not considered to be a good practice, which would result in cheap quality. But with time, due to multiple factors, the spent (waste) refractory utilization has got importance, priority, and momentum. For about last two decades, the work on waste refractory utilization has witnessed increased attention, developmental works, and commercial activity. The primary driver for waste refractory utilization is the movement for a "zero-waste" culture across the industries, including the refractory users and manufacturers.

At the early stage, recycling of waste refractories has not received proper attention primarily due to the wide abundance of low-cost virgin raw materials, low disposal costs as refractories are by and large inert in character, and less awareness on environmental issues. On the other hand, the expenditure on refractories and its waste generation was mainly dealt with by reducing the specific refractory consumption value. But, the scenario started changing in the late of last century, primarily due to the following factors:

1. Rapid increase in the prices of virgin raw materials, both natural and synthetic
2. Decrease in the availability and the supply of raw materials, especially imported ones
3. Increase in the duties and taxes of the exporting countries
4. Increase in the transportation charges mainly due to increased fuel cost
5. Increased pressure from government on environmental issues and carbon footprint
6. Implementation of higher charges on landfill and emissions
7. Tax incentives by government to motivate companies to increase waste recycling
8. Economic recession and high competition among the refractory manufacturers, forcing them to reduce input cost to sustain in the market

All these above factors along with various changes in the refractory manufacturing and user industries that have occurred with time have caused a shift to the use of reclaimed refractories. As on today, waste utilization and recycling have become a major buzz word, as we all have become aware of global warming and climate change. The refractory industries can greatly contribute in conservation of the environment by increasing the recycling of reclaimed refractory materials generated from its use in different user industries. Not only the environmental issues, but recycling of refractories also provide a range of ecological and cost advantages to the industry, as mentioned below:

1. Amount of disposed refractory is greatly reduced, thus reducing the disposal cost and reduction in landfill materials.
2. Better ecological balance due to lesser land fill.
3. Lesser impact on the earth's crust and its fertility with the conservation of mineral resource.
4. Prefered material reduces firings like calcination, dead burning, etc., saves energy and fuels, reduces refractory manufacturing process steps, improves economy and efficiency, and also reduces the carbon footprint.
5. Imported, costly raw materials are substituted.
6. Generation of work for the local communities.

7. Reduction in refractory manufacturing cost.
8. Reduced stock holding, especially for the imported raw materials.

Currently, the global refractory production is in the tune of 50 million tons per year, wherein two-thirds of the production is being consumed by iron and steel industries. Hence, the main waste refractory generator is iron and steel industry, only with close to similar extent of utilization. Again, during iron and steelmaking process, refractories get corroded, abraded, and worn away or get dissolved in slag. This dissolution of refractory is around 30%–40% of the total refractory used, and similar dissolution/wear away is common for all the refractory application areas. Hence, on an average, the global waste refractory generation is in the tune of 30–35 million tons per year, which is very high and needs proper planning and attention to control the same.

14.4.1 RECYCLING PROCESSES

There are two main types of recycling processes for any material, namely, open-loop recycling and closed-loop recycling. These processes differ in the overall sustainability of the supply chain of the raw materials processed. In refractory also, both the types exist and differ greatly in the utilization of the waste refractory.

14.4.1.1 Open-Loop Recycling

Open-loop recycling is a process that converts the recycled material into both new raw materials (which can be used as production inputs) and waste products. Typically, the recycled material developed is used for purposes different from their former—the pre-recycled one. The input into the recycling process converts the original material to a new raw material that can be used as an input into another completely different process. This change in character of the material generally occurs due to heat, chemical reactions, and physical change, which is often associated with a degradation of the material being recycled.

One of the most common open-loop recycling of refractory application for spent refractory is its use as road-base aggregates. Another classic example for the open-loop recycling is the use of basic refractories, like, doloma and magnesia, as slag former or conditioner in metallurgical processes. Such a use increases the MgO saturation level of the slag and reduces refractory (MgO–C) attack, thus increasing the lining life. The use of spent refractories as slag conditioner has also advantages like energy savings, savings in fluxes, reduction in landfilling activities, etc.

14.4.1.2 Closed-Loop Recycling

Closed-loop recycling, on the other hand, is a recycling process in which the waste material is recycled back into itself or a similar product without significant degradation or waste generation. It focuses on the supply chain sustainability. In this recycling process, closed-loop systems are important, as it targets to recycle all the materials back to the same or similar-type product without any (significant) loss of material, providing greater economic values. This recycling process is of greater importance for refractories, as it restricts the use of virgin materials leading to significant energy and associated greenhouse gas emissions savings. For magnesia (MgO) refractory manufacturing, use of spent magnesia helps to remove the energy-intensive calcination or dead burning process of the primary raw material magnesite ($MgCO_3$), and thus saves an energy in the range of 6–12 GJ per ton of MgO.

However, use of spent refractory may affect the physicochemical properties of the developed refractory compared to that of the only virgin raw material-based ones. Spent refractories have faced the high-temperature processing, were in contact with the metal, slag, glass, or other process materials, and got contaminated during its lifetime. Hence, there is an increase in the amount of impurities like CaO, Fe_2O_3, SiO_2, etc. in the recycled refractories. Greater presence of the impurities deteriorates the refractory properties, like density, strength, chemical resistance, etc., and the deterioration increases with the increasing amount of spent material used. Also, use of spent material decreases the durability of the developed refractory due to low-melting-compounds formation. For magnesia-based refractories, the liquid phase is in the $MgO–CaO–Fe_2O_3–SiO_2$, and for alumina based refractories, it is in the $Al_2O_3–CaO–Fe_2O_3–SiO_2$ system. Thus, the extent of use of recycled material in the developed refractory is limited and dependent on the purity of the recycled fraction. Hence, proper processing of the recycled portion is essential to get good-quality spent material.

14.4.2 MAJOR STEPS INVOLVED IN REFRACTORY RECYCLING PROCESS

The waste refractories can be utilized as the raw material for making new refractory and can provide all the required characteristics for the new refractory. The primary requirement for such uses are dependent on chemical composition (chemistry must not vary widely) and quality assurance (uniformity in properties with time). To satisfy these requirements, it is important for the recycled refractories to be sorted and pretreated correctly before use and provide suitable properties in the newly developed refractory. Hence, primarily sorting and pretreatment processes are the essential steps for utilization of waste refractory.

14.4.2.1 Sorting

Traditionally, sorting is done manually, and proper attention is essential, as different types of refractories are mixed in the waste yard, and they are commonly contaminated with the process materials, like, metal, slag, glass, etc., as per the source industry. Manual sorting of waste refractories is primarily depended on operators' expertise. Knowledge on the common co-occurring refractories and their physical characteristics are important for sorting. Operators generally separate the waste refractories based on appearance, color, density, etc. However, the process may not be highly reliable due to the presence of dust and contaminants that restricts visual identification. Also, the manual sorting process is time consuming and less reliable due to poor dump yard environment, weight of broken pieces, poor wages, etc.

There are also possibilities of inline automated sorting system based on computerized camera working on grayscale imaging (or by using Charge-coupled device) for color measurements. However, change in color due to dust, abrasion, and contamination can commonly result in erroneous sorting process. Sorting based on chemical composition is a better and reliable option; however, surface composition of the waste refractories may vary due to contamination. Also, analyzing the chemistry (say, X-ray fluorescence technique) of each and individual waste refractory pieces for sorting is impractical and not feasible commercially. Further advanced techniques are being worked out for accurate and faster sorting process, namely, electrodynamic fragmentation technique (separates compounds selectively by using pulsed power discharge), laser-induced breakdown spectroscopy (based on element-specific spectral emission lines to identify the pieces), etc. are being tried in different studies, but their commercial feasibility, economy, and success are yet to be established.

14.4.2.2 Pretreatment

This is the second major step for refractory recycling process and is more instrument based (automated) one. Pretreatment mainly aims for the purification of the presorted fractions. To separate the desired refractory portions from the contaminations and impurities, like metal, slag, glass, cement clinker, etc. the sorted refractory pieces are commonly processed for operations like, crushing, grinding/milling, screening, magnetic separation, and color separation. Light-weight contaminants, like dust, paper, plastics, etc., present in a very minute extent but highly harmful, can be separated by common air classifiers that are based on upward vacuum extraction technique.

Screening is done to purify the spent refractory using different sieves, if the contaminants are present in a specific size fraction. Similarly, magnetic separation is important if the contaminants are magnetic in nature. This is more useful for waste refractories from iron and steel plants, and commonly band or drum magnets are used for separation purpose. For carbon-containing refractories, the carbon is commonly burnt out to get the oxide component of the refractory. Also, there are patented processes based on flotation, acid treatment, washing to selectively recover carbon and oxide components separately.

SUMMARY OF THE CHAPTER

Most of the scientific and technological inventions and developments that are known today would not have been successful if proper support from refractories were not available. The existence of nearly all materials that we see around us or use in everyday life are directly or indirectly dependent on refractories. Refractories are the facilitating or enabling material, in some way, for any development, and are essential for any high-temperature industry or process.

The refractory technology is being developed since the beginning of the civilization and is upgrading itself with the progress of time. It has developed as a diversified field of technology, which once used to be considered as a mere support material for high-temperature processing, grew into a comprehensive system with heat insulation, process control, productivity, and efficiency of the user industries. Remarkable advancements in refractory were seen during the periods of the Industrial Revolution and world wars.

In recent times, use of nanotechnology has gained significant attention in the field of refractory research. Nanotechnology is becoming a prime direction of work in refractory research and development. Interest in the use of nanoparticles in refractory has grown due to the possibility of a marked improvement in properties of normal substances by using or converting them to a nanometric dimensions. Use of nano-technology, in terms of nanoparticles, nanomaterial, nanoadditives, and nanostruc-tured materials, is the most common activity for refractories. Use of different types and amounts of nanomaterials (oxides and nonoxides) in refractories and control of microstructure for properties development even at much lower temperatures are the prime attention for the refractory technologists.

Refractories also impact greatly on the environment. The energy load of any high-temperature process is quite large. Most of the environmental impact of any high- tem-perature operation is associated with the CO_2 generation for the required high energy load. Reduction in required energy load has multiple folds of benefits, and refractories plays the biggest role. Also, the refractory production and practice have impact on the environment. For a better environment, it is essential to use ecofriendly manufacturing process with nonhazardous raw materials and additives; reduce the operating tempera-ture of processing; use of proper dust collector, heat recovery system; restrict the use of health-hazardous chromium-bearing materials, ceramic fibers, etc.

Refractories perform their duties throughout the service life, and after that they become unsuitable to perform and are replaced. The used refractories become waste, and utilization of the waste refractory is becoming essential for better environment and economy. Recycling has become a major buzz word all over the world mainly due to global warming and climate change. High-temperature processing industries can greatly contribute in conserving the environment by increasing the recycling of reclaimed refractories generated from operations.

Refractories are made up of high-quality raw materials, mostly natural, whose quality are degrading with time and prices are rising. Scarcity of good-quality raw material along with environmental issue and disposal cost of waste refracto-ries have encouraged the technologists for recycling of refractories. It is only in this century that recycling in new refractories has gained interest due to economic ben-efits (cheaper raw materials, lower treatment costs) and environmental issues (lower energy demand and CO_2 emissions compared to virgin materials).

Though the technical feasibility of such reuse has been studied for many refracto-ries but has been applied in only limited types. The main bottleneck for recycling is the quality assurance of the recycled fractions. Sorting is the major hurdle for recy-cling, as manual sorting is the only commercially feasible option in industry, which is error prone and requires a prior knowledge. Also, the mindset of the refractory users need to be changed to accept refractories containing recycled fractions.

QUESTIONS AND ASSIGNMENTS

1. Describe in your words about the progress of refractories with time.
2. Why nanosized materials are becoming important for technical field?
3. Mention few examples of use of nanotechnology in refractory field.
4. How refractories can impact the environment?
5. How monolithics can help in keeping better environment?
6. Discuss few actions taken by refractory industry for a safe and better environment.

7. What are open-loop and closed-loop recycling processes?
8. What are the advantages of recycling of refractories?

BIBLIOGRAPHY

1. T. Emit, Future outlook of refractories for iron and steelmaking, *Journal of the Korean Ceramic Society* 40 [12] 1141–1149 (2003).
2. M. A. Deneen and A. C. Gross, Refractory materials: the global market, *The Global Industry, Business Economics*, 45, 288–295 (2010).
3. T. Takeuchi and N. Taki, Progress and future prospects of refractory technology of nippon steel corporation, *Nippon Steel Technical Report*, 125 3–9 (2020).
4. Y. Tang, Y. Shi, Y.Li, X. Yuan, R. Mu, Q. Wang, Q. Ma, J. Hong, S. Cao, J. Zuo and J. Kellett, Environmental and economic impact assessment of the alumina–carbon refractory production in China, *Clean Technologies and Environmental Policy*, 21, 1723–1737 (2019).
5. R. Sarkar, Nanotechnology in Refractory Castable: an overview, *InterCeram: International Ceramic Review*, 67 (S1) 22–31 (2018).
6. V. Pilli and R. Sarkar, Nanocarbon containing Al_2O_3 – C continuous casting refractories: effect of graphite content, *Journal of Alloys and Compounds*, 735, 1730–1736 (2018).
7. V. Pilli and R. Sarkar, Study on the nanocarbon containing Al_2O_3 – C continuous casting refractories with reduced fixed carbon content, *Journal of Alloys and Compounds* 781 149–158 (2019).
8. A. K. Singh and R. Sarkar, Nano mullite bonded refractory castable composition for high temperature applications, *Ceramics International*, 42 [11] 12937–12945 (2016).
9. S. Behera and R. Sarkar, Nano carbon containing low carbon magnesia carbon refractory: an overview, *Protection of Metals and Physical Chemistry of Surfaces*, 52 [3] 467–474 (2016).
10. M. Bag, S. Adak and R. Sarkar, Study on low carbon containing MgO-C refractory: use of nano carbon, *Ceramics International*, 38 [3] 2339–2346 (2012).
11. M. Bag, S. Adak and R. Sarkar, Nano carbon containing MgO-C refractory: effect of graphite content, *Ceramics International*, 38 [6] 4909–4914 (2012).
12. L. Horckmansa, P. Nielsena, P. Dierckxa and A. Ducastelc, Recycling of refractory bricks used in basic steelmaking: a review, *Resources, Conservation & Recycling*, 140 297–304 (2019).
13. Z. Ningsheng and L, Jiehua, Green refractories – Concepts, approaches and practices, *Refractories World Forum*, 8 [3] 99–110 (2016).
14. D. V. Kuznetsov, D. V. Lysov, A. A. Nemtinov, A. S. Shaleiko and V. A. Korolkov, Nanomaterials in refractory technology, *Refractories and Industrial Ceramics*, 51 [2] 61–63 (2010).
15. F. G. Simon, B. Adamczyk and G. Kley, Refractory materials from waste, *Materials Transactions*, 44 [7] 1251–1254 (2003).

Index

Note: **Bold** page numbers refer to tables; *italic* page numbers refer to figures.

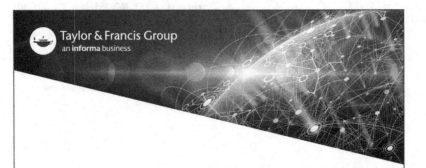

Taylor & Francis eBooks

www.taylorfrancis.com

A single destination for eBooks from Taylor & Francis
with increased functionality and an improved user
experience to meet the needs of our customers.

90,000+ eBooks of award-winning academic content in
Humanities, Social Science, Science, Technology, Engineering,
and Medical written by a global network of editors and authors.

TAYLOR & FRANCIS EBOOKS OFFERS:

A streamlined
experience for
our library
customers

A single point
of discovery
for all of our
eBook content

Improved
search and
discovery of
content at both
book and
chapter level

REQUEST A FREE TRIAL
support@taylorfrancis.com

 Routledge
Taylor & Francis Group

 CRC Press
Taylor & Francis Group